QUANTUM
MECHANICS

Its Early Development and the Road to Entanglement and Beyond

NEW ENLARGED EDITION

QUANTUM MECHANICS

Its Early Development and the Road to Entanglement and Beyond

NEW ENLARGED EDITION

Edward G. Steward

Emeritus Professor, City University, London

with a contribution by Sara M. McMurry

Imperial College Press

ICP

Published by

Imperial College Press
57 Shelton Street
Covent Garden
London WC2H 9HE

Distributed by

World Scientific Publishing Co. Pte. Ltd.
5 Toh Tuck Link, Singapore 596224
USA office: 27 Warren Street, Suite 401-402, Hackensack, NJ 07601
UK office: 57 Shelton Street, Covent Garden, London WC2H 9HE

British Library Cataloguing-in-Publication Data
A catalogue record for this book is available from the British Library.

QUANTUM MECHANICS
Its Early Development and the Road to Entanglement and Beyond
(2nd Edition)

ISBN-13 978-1-84816-769-8
ISBN-10 1-84816-769-5
ISBN-13 978-1-84816-770-4 (pbk)
ISBN-10 1-84816-770-9 (pbk)

Typeset by Stallion Press
Email: enquiries@stallionpress.com

Printed in Singapore.

"To not lose the wood among the trees"
after G. K. Chesterton *The Wise Men*

Dedicated to
the memory of hearing lectures by
Niels Bohr, Max Born, W. L. Bragg, H. T. Flint,
Max von Laue and W. Wilson

Contents

Acknowledgements

For encouragement to write a book of this type, I thank Spencer Weart (Center for History of Physics, American Institute of Physics), and David Cassidy (author of *Uncertainty: The Life and Science of Werner Heisenberg*).

It has been a great pleasure to have the final chapter contributed by Dr Sara McMurry (Trinity College, Dublin), author of a successful textbook on quantum mechanics. Sara has also critically read Chaps. 7 and 9, spotted errors and made helpful suggestions.

Professor Gordon Davies, Head of the Physics Department at King's College, London, where I was originally a student, has given me much encouragement, has kindly read Chaps. 1 to 9 and the appendices, and has given me valuable comments and suggestions.

Apart from the original research publications, I have consulted many of the books and articles listed in the Bibliography. In some instances, I have contacted the authors and I would particularly like to express my gratitude to Jeremy Bernstein, David Cassidy, William Cropper, Ian Duck and Clayton Gearhart.

Finally, but most importantly, it gives me very special pleasure to express my gratitude to my son Martin at the University of Manchester for enabling me in so many ways to achieve my goal.

Preface to the First Edition

This book provides the reader with the unique combination of an explanation of the origin and establishment of quantum mechanics, with the mathematics in a digestible form, together with a descriptive survey of the later developments up to the present day. In addition to its importance for those studying physics, it is also a valuable treatment for those studying the history of science.

The mathematical treatments closely follow the original, but in modern terms, with uniform symbolism throughout as far as possible, and with simplifications (e.g. the use of one dimension instead of three) to avoid visual indigestion with unnecessarily complicated and over-decorated mathematics. Appendices include notes on topics where it is essential to recall and have in mind certain details that are directly relevant to the subject of the book. There is also an appendix giving brief biographical notes on the central characters.

The treatment fills the gap that exists between (i) the 'student' texts where, of necessity, little attention is paid to the historical development and the aim is more directed to applications, (ii) the more advanced texts that assume much and are intended for those becoming deeply involved in the subject, the basics needed often being either assumed or derived as a matter of expediency in different ways, with hindsight, from the original, and (iii) historical /biographical accounts.

We have recently celebrated the centenary anniversaries of the birth of the quantum in 1900 and the birth of the Special Theory of Relativity in 1905. Both were major advances that profoundly changed our picture of the physical world. The first changed our understanding of the nature of electromagnetic radiation and matter on the sub-microscopic scale, and the second changed our concept of space and time on the cosmic scale.

After an introductory chapter setting the scene for the book, there is a chapter outlining the way the need arose for Einstein's Special Theory of Relativity. As it is not the subject of the book, only those details that are involved are introduced and explained. A chapter on the arrival of Planck's 'energy elements' is then followed by Einstein's work at that time including his concept of light as 'light quanta', Bohr's extension of that to the atom and its rôle in understanding optical spectra, de Broglie's notion of wave–particle duality applying to matter, and thence to Schrödinger's wave mechanics, and Heisenberg's matrix mechanics. The conflict that had increasingly become evident between the different theories is then the subject of a chapter that explains how this led to the notion of Complementarity, to Heisenberg's Uncertainty Principle, and the 'Copenhagen Interpretation'. The aftermath of that and the concept of 'entanglement', and other developments, are described in the final chapter.

Special mention needs to be made of the inclusion of a chapter devoted to matrix mechanics. It is often described as outside the scope of introductory texts because it is too difficult, and where it is included it is usually dealt with inadequately and inaccurately. For example, some authors state that it was a method based on matrix algebra though this was not recognised as such at that time, others state that it was based on Fourier analysis, and yet others state that it was based on Bohr's Correspondence Principle. Individually, these are only part of the truth, and unless the overall picture of the actual course of events is appreciated, the reader cannot fully understand what is involved or how the subject was originally developed. Also included here is a simple treatment of how matrix mechanics and wave mechanics are equivalent formulations.

For some years, I have been distilling out the main thread of the development of the subject from the original literature and the innumerable historical accounts and biographies. It is justifiably open to the criticism that science does not proceed along a single linear sequence of contributions but by a much more subtle combination of human factors and cross-fertilization of ideas. Nevertheless, the actual outcome of what it amounted to is the sequence recounted here in an accessible way. With references to the original works, to reviews and biographies, and a useful bibliography, the reader is then uniquely well-equipped to go further into the subject and appreciate its more subtle details.

Preface to the Second Edition

The purpose of this book remains unchanged, namely to combine an accessible explanation of the origins and early development of quantum mechanics with a survey of the subsequent developments.

The evident usefulness of the first edition has called for an updating and broadening of the coverage of the book. To that end, Sara McMurry has expertly extended the coverage of the chapter on Indeterminacy and Entanglement with a new chapter entitled 'Interpretations of Quantum Mechanics', which surveys a wide range of current topics. These include the multiverse, 't Hooft's ideas for a deterministic local field theory, a summary of the de Broglie–Bohm pilot-wave theory and Anthony Valentini's development of it, and speculative concluding comments on the way ahead.

In view of the continuing relevance of de Broglie's pioneering work, a chapter has been added in the form of 'A Reflective Interlude', to look in more detail at the origin and early years of wave–particle duality, with emphasis on trying to discover, as far as possible, what was the physical reality implied by his work as it progressed. This task has mainly involved revisiting the original publications, but also the series edited by Mehra and Rechenberg, and the writings of Max Jammer, Georges Lochak (de Broglie's biographer and one-time colleague) and many others. It has also been greatly helped by Bacciagaluppi and Valentini's recent work in assembling, for the first time, a comprehensive account of the historically important 5th Solvay Conference and its background and aftermath.

Finally, an appendix has been added, dealing with the origin of the fascinating concept of Planck Units. Originally no more than the result of a search for 'natural units', Planck Units and Planck Space have become relevant in topics of theoretical physics ranging from General Relativity to String Theory.

The opportunity has also been taken to make corrections and minor amendments to the original text. References to the original publications, Bibliography, and Further Reading have been updated, where appropriate, to continue to enable the reader to explore the various topics in greater depth.

Again, it gives me great pleasure to record my thanks to Sara for her enthusiastic collaboration, and to my son Martin for help with some aspects of preparing the manuscript.

Edward Steward

About the Author

Edward Steward graduated at King's College, London, and joined the GEC Research Laboratories where he was engaged in X-ray diffraction studies of a wide variety of materials. He gained his PhD, became a Special Visiting Lecturer at Sir John Cass College, and left GEC to become a Reader in the Physics Department at Northampton College of Advanced Technology in London, subsequently City University.

Now retired, Professor Steward's research at City University, for which he was awarded a DSc (London), included the combined use of X-ray diffraction and quantum mechanics in the study of drug-receptor interactions.

His professional activities have included membership of various committees of the Institute of Physics in the UK and Commissions of the International Union of Crystallography.

Author of over 100 publications and contributing author to a number of books and two educational films, this book can be seen as complementary to his well-known book on Fourier Optics.

Chapter 1

Setting the Scene

1.1. Introduction

Quantum Mechanics is commonly defined as the system of mechanics that was developed from Quantum Theory to explain the properties of atoms and molecules. **Quantum Theory** itself had earlier been established as a consequence of Planck's introduction in 1900 of the concept of '**energy elements**' (named '**energy quanta**' by Einstein in 1905(a)) to resolve the inability of Classical Physics to account fully for the spectrum of radiation from a hot body ('**black-body radiation**' explained below). However, the concept of 'energy elements' could equally well have been arrived at, and probably more easily, in resolving other problems confronting Classical Physics at the end of the 19th century — for example the failure to explain the **specific heat of solids** at low temperatures, or why in the **photoelectric effect** the energy of ejected photoelectrons depends only on the frequency of the incident light and not its intensity.

A number of developments led to the establishment of a 'quantum mechanics'. Firstly, Planck's discovery immediately overturned the universally accepted notion in Classical Physics that energy is a continuous variable. Instead, it is 'granular' and 'discrete', to quote descriptions often used. That concept was taken forward crucially in 1905 by Einstein, who explained details in the **photoelectric effect** by proposing that radiation itself is 'quantised', a concept alien to Planck who only saw its *generation* as involving elemental energy units.

1

Then, in 1913, Bohr showed the role of the quantum concept in explaining the stability of Rutherford's 1911 planetary model of the atom, in terms of '**stationary states**'. In 1923 came de Broglie's suggestion that the concept of **wave–particle duality** of light, resulting from Einstein's quantum explanation of the photoelectric effect, also applies to matter particles. In the hands of Schrödinger, Heisenberg, Born, Dirac and others, '**quantum mechanics**' was born. These form the subject of Chaps. 7 and 8.

It is fascinating to realise that the basic discoveries by Planck, Einstein and Bohr sprang from earlier studies of the radiation from the Sun and the colour of flames. It was largely the work on those topics by Kirchhoff in 1859–1860 that provided the fertile ground. Sections 1.2 and 1.3 briefly outline how that came about. (Kirchhoff was already known for his laws in 1845 concerning electrical networks — work foreshadowing Maxwell's 1864 Electromagnetic Theory of radiation.)

In §1.4, we set the scene for Planck's work on tackling the problem mentioned above concerning the explanation of the details of the observed spectrum of heat radiation — the subject of Chap. 3. Then, in §1.5 we note the problems concerning the nature of light, and indicate how this led to Einstein's **Special Theory of Relativity**. This is the subject of Chap. 2, and it played a major role in the development of the Quantum Theory and Quantum Mechanics. Finally, in §1.6, we briefly note the range of outstanding contributions made by Einstein, referred to in various parts of this book, and the arrival of the concepts of **Complementarity**, the **Uncertainty Principle**, the **Copenhagen Interpretation**, and '**entanglement**', that are the subjects of Chaps. 9 and 10.

1.2. Light and Heat: Kirchhoff's 'Black-body Radiation'

Key steps leading to the way in which Kirchhoff's work revealed the problem that Planck was to solve can be summarised as follows:

1800 Herschel showed that heat radiation (e.g. the heat of sunlight) is the extension into the infrared, i.e. beyond the red end of the visible light spectrum. He did this by using a prism and moving a thermometer along the spectrum produced by sunlight.

1802 Wollaston showed that the solar spectrum contains 'dark' lines.

1822 Fraunhofer had by now measured the dark lines using a diffraction grating. He found that the strongest (a doublet) was at the same position as the yellow light of many flames.

1826 Fox Talbot studied the colours of flames containing different substances, showing the potential for the use of spectroscopy in chemical analysis.

1833 W. H. Miller suggested that the Fraunhofer dark lines in the solar spectrum were due to absorption in the outer layers of the Sun.

1833 Ritchie demonstrated experimentally, in an elegantly simple way (described in elementary textbooks), the proportionality of the heat-emitting and heat-absorbing powers of a surface.

1858 Balfour Stewart showed that a radiating body in equilibrium at a given temperature emits and absorbs radiation at equal rates (Prévost's 1791 Theory of Exchanges) applies at all wavelengths, i.e. to light as well as heat radiation.

1859 Kirchhoff, independently of Balfour Stewart, and more rigorously, came to the same conclusion. In Kirchhoff's classic paper presented in 1859, entitled 'On the relation between the emission and absorption of light and heat', he used thermodynamic principles to show that "for rays of the same wavelength the ratio of the emissive power to the absorptivity, at the same temperature, is the same for all bodies". This was subsequently known as 'Kirchhoff's Law', and more rigorous proofs followed.

It was in a second paper that Kirchhoff introduced the notion of what he called a 'black-body', defining it as a body that absorbs all the radiation incident upon it. Together with the theorem, it meant that with such a 'black-body' one could have an equilibrium spectral distribution independent of everything except the temperature. This was clearly of enormous fundamental significance.

A close approximation to a black-body was achieved by constructing a cavity with highly-insulating walls and closed except for a very small orifice. Experimental studies of the radiation were made and it became increasingly evident that the details could not be explained by classical methods. That is where evidence of the quantum nature of energy became apparent, and that is the subject of Chap. 3.

1.3. Kirchhoff's Work on Optical Spectra

Not only had Kirchhoff's concept of black-body radiation led to Planck's discovery of the '**energy quantum**', but in his work leading to the formulation of his theorem he had explored in more detail Fraunhofer's observations that the dark lines in the solar spectrum coincided with the bright lines in the spectra of flames. He, and others, were able to show that chemical elements have unique, characteristic spectra, and that the emission and absorption frequencies are the same for a given element — this became known as Kirchhoff's '**law of radiation**'. The solar dark line (a doublet) was identified with the D-line doublet of sodium.

This aspect of Kirchhoff's work led to the other major discovery of the role of the quantum concept. It came about as follows.

With Bunsen, Kirchhoff studied the solar spectrum in more detail and by 1860 he was able to identify the chemical elements in the atmosphere of the Sun. This was the foundation of chemical spectroscopy and it also led to a new era in astronomy. However, problems lay ahead.

As an extension of Kirchhoff's work, Balmer produced in 1885 an empirical formula for the hydrogen spectrum, but a scientific basis was lacking. In 1911, Rutherford's planetary model of the atom was deduced from experiments by Geiger and Marsden, at the suggestion of Rutherford, on the scattering of α-particles by a thin gold foil. However, the stability of such an orbital system, in which orbital electrons would, classically, be expected to spiral into the nucleus, was unexplained. In 1913, Bohr showed that if one introduced the concept of orbital 'stationary states' between which discrete, quantised, transitions occur, then the observed spectral lines could be explained. But what was the nature of the '**stationary states**'? The way in which that was resolved is the subject of Chap. 5, with further developments in Chap. 7.

1.4. Planck's Route to Tackling the Black-body Radiation Problem

Planck's approach to dealing with the black-body radiation problem was by a circuitous route.

In his formative years as a physicist he was influenced very much by the writings of Clausius, especially with regard to the Second Law of Thermodynamics. Based on experimental observations, the Second Law was variously expressed, but to the effect that it is impossible to cause heat to pass from one body to another at a higher temperature without the expenditure of mechanical work. Then, in 1865, Clausius introduced the concept of entropy and reformulated the Second Law to state that the entropy of an isolated system always increases or remains constant (see Appendix A), i.e. changes are irreversible.

Strongly believing in the absolute validity of the Second Law, Planck was opposed to the statistics approach that Boltzmann took to thermal equilibrium, in which Boltzmann was following Maxwell's earlier view of the Second Law as statistical in nature, and which did not exclude the possibility of reversibility. Planck's opposition to the idea of reversibility and his preference for **Classical Thermodynamics** was going to influence him when his attention turned to the black-body radiation problem. However, as we shall see in Chap. 3, he found that he had to change reluctantly to a '**probabilistic**' approach, though not using Boltzmann statistics.

Another topic on which Planck had worked prior to tackling the black-body problem, and which was also going to influence him, was triggered by his learning of Hertz's discovery in 1887 of the electromagnetic waves emitted by oscillating charges (seen as an important confirmation of Maxwell's electromagnetic theory). It led Planck to visualise black-body radiation as originating in sub-microscopic **linear harmonic oscillators**, or '**resonators**' (the simplest body that can absorb and emit radiation) in the walls of a black-body cavity as conceived by Kirchhoff (which involved no stipulation about the nature of the cavity wall). It led him to derive the following expression relating the black-body radiation frequency distribution, $\rho(v, T)$, to the average equilibrium energy, $E(v, T)$, of a (damped) harmonic oscillator of frequency v at temperature T in a cavity wall:

$$\rho(v, T) = \frac{8\pi v^2}{c^3} E(v, T). \tag{1.1}$$

This 'theorem' was to form the basis of much that followed and it played an important role in Einstein's subsequent studies.

To determine $\rho(v, T)$, it was of course necessary to determine $E(v, T)$ and that is where the difficulties arose. Suffice it to say here that using the **Equipartition of Energy Principle** of Classical Physics led to a distribution that was not in accord with experimental data.

With the resonators that Planck envisaged, quite legitimately as we have noted, as a model for the emitters/absorbers in the walls of a black-body cavity, and his strong belief in the Second Law of thermodynamics, he embarked on tackling the problem that was arising in finding a scientific basis for the experimental measurements of black-body radiation. That is the fascinating story presented in Chap. 3, leading to the birth of the quantum. Before that, we need to recall some relevant aspects concerning the state of knowledge, at that time, about electromagnetic radiation itself.

1.5. Light and the Aether

The transverse wave model of light was established by Young's **double-aperture experiment** of 1801, and elaborated by the work of Fresnel around 1815.

Maxwell's 1864 electromagnetic theory then suggested that the wavemotion of light consists of oscillating electric and magnetic fields travelling through an 'aether', a medium through which the Earth moves, as suggested by Bradley's observations in 1726 of the 'stellar aberration of light'.

There were various conflicting experiments and theories to resolve the question of the existence of the aether. The famous **Michelson–Morley experiments**, during 1881–1887, failed to reveal any evidence of the Earth moving through an aether, and this could only mean that the aether (if it existed) was travelling with the Earth. However, that was at variance with e.g. Bradley's observed stellar aberration.

The problem continued, with much discussion and theorising, but it was not until 1905 that a solution was found by Einstein with his **Special Theory of Relativity**, and later it was experimentally verified. Not the subject of this book, in Chap. 2 we recount just the details that we shall need because of their vital role in the development of the quantum theory.

1.6. Post-1900

Einstein's contributions to so many aspects of the development of quantum physics started at the beginning of the 20th century.

His first two papers were published during 1901–1902 and were concerned with the phenomenon of capillarity, and molecular forces. Then, during 1902–1904, he published three papers in which he laid down his own foundations of **statistical physics**. This topic was to play an important role in the development of the quantum theory and it occupied him for the next quarter of a century.

The year 1904 also saw Einstein's first reference to the **black-body radiation** work, where he noted Planck's use of the statistical interpretation of the **entropy of ideal gases**. That marked the start of Einstein's interest in the quantum theory in the context of statistics and led him to a study of **energy fluctuations** in thermal equilibrium, and related topics. Because this latter work became involved in different ways in the developments that followed, we devote Chap. 4 to looking at the main principles so that we can draw on them as necessary without breaking the main flow. A feature that will strike the reader is the way in which radiation was treated as a 'quantum gas' at various stages throughout.

After Planck's introduction of the quantum in the form of his 'energy elements' in 1900, Einstein extended the concept in 1905, in a paper concerned with the photoelectric effect, to apply to light itself, with light envisaged as travelling as '**light quanta**', later to be known as '**photons**'.

That was just one of five papers that Einstein produced in 1905, five papers that, as some have rightly put it, 'shook the world'. Produced in his spare time as a technical expert in the Patent Office in Bern, it is also remarkable that the five papers were not based directly on experiment or sophisticated mathematics. As Gerard 't Hooft — who shared with Mortinus J. G. Veitman the 1999 Nobel Prize for Physics for their work on quantum theory — has said, "Einstein's work stands out not because it was difficult but because nobody at that time had been thinking the way he did."

The five papers comprised the one mentioned above, on the photoelectric effect and the light quanta hypothesis, two on **Brownian Motion,** and two on his **Special Theory of Relativity.**

Einstein's concept of light travelling as discrete quanta was of course in conflict with the longstanding picture of light as a wave-motion, as demonstrated by diffraction and interference effects and extensively studied by Young and Fresnel. In 1923, de Broglie suggested that such wave–particle duality might also apply to matter particles (confirmed experimentally). This was developed by Schrödinger and others to become '**Wave Mechanics**', the subject of Chap. 7.

At virtually the same time, Bohr's work on atomic spectra, referred to in §1.3, was drawing Heisenberg's attention to the question of the reality of the 'position', 'orbit', 'motion', etc. of an electron, and the problem of observing them experimentally. This led Heisenberg to propose in 1925–1926 a new mechanics in which such unobservables are not involved. It was rapidly developed by others and became the '**Matrix Mechanics**' of today. That is the story recounted in Chap. 8.

These two forms of '**Quantum Mechanics**' were soon shown to be mathematically equivalent, though based on different conceptual models.

To deal with wave–particle duality, Bohr developed his **Principle of Complementarity.** Also at that time, Heisenberg's **Uncertainty Principle** emerged as a consequence of matrix mechanics. Attempts to reconcile all the different models and interpretations eventually led to what became known as the '**Copenhagen Interpretation**'.

It was challenged on several occasions by Einstein, who never abandoned his strong belief in continuity and **causality** (the direct connection between cause and effect) — thus his much quoted remark that "God does not play dice". In the following years, the Copenhagen Interpretation and its implications were studied and tested by many and the notion of '**entanglement**', first mooted by Schrödinger, was developed. These are the topics of Chaps. 9 and 10.

The outline chronology is as follows:

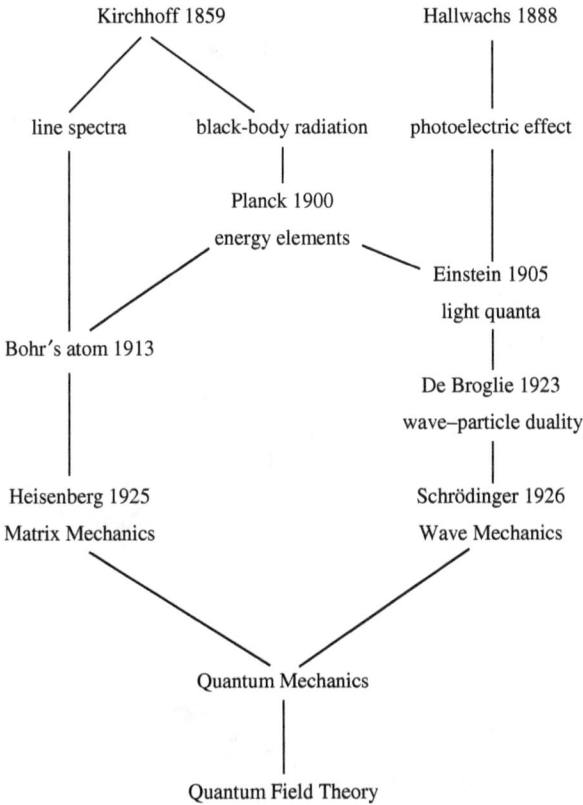

```
        Kirchhoff 1859                        Hallwachs 1888
          /      \                                  |
         /        \                                 |
  line spectra   black-body radiation        photoelectric effect
       |                |                            |
       |           Planck 1900                       |
       |          energy elements                    |
       |              /        \                      |
       |             /          \               Einstein 1905
       |            /             \             light quanta
  Bohr's atom 1913                                   |
       |                                       De Broglie 1923
       |                                      wave–particle duality
       |                                             |
  Heisenberg 1925                            Schrödinger 1926
  Matrix Mechanics                            Wave Mechanics
          \                                        /
           \                                      /
            \                                    /
             \          Quantum Mechanics       /
              \              |                  /
               \             |                 /
                      Quantum Field Theory
```

Einstein remained convinced that underlying his work, and that of others, based on statistical probabilities, though giving good results in applications, there is nevertheless a fundamental 'complete theory' where continuity and causality lie. He spent his remaining years in an

unsuccessful quest for a **Unified Theory** — an extension of his General
Theory of Relativity to include Quantum Mechanics — and which,
despite more recent developments, continues to be elusive.

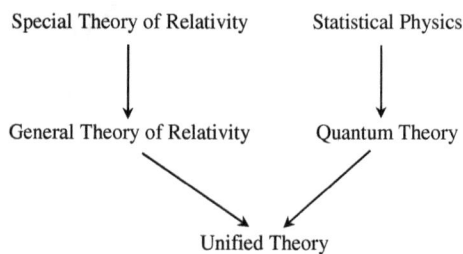

Special Theory of Relativity   Statistical Physics

General Theory of Relativity   Quantum Theory

Unified Theory

Chapter 2

Light: The 'Aether' and the Special Theory of Relativity

2.1. Introduction

Theories about the nature of light go back to the Greek philosophers and beyond. It is a complex story of conflicting evidence and changing theories, and there continue to be unresolved problems.

Newton's **corpuscular theory of light** held sway to the end of the 18th century. He was impressed by the fact that, unlike waves, light travels in straight lines, and the first of his **laws of motion** (1687) explained motion in straight lines. However, to explain simultaneous refraction and reflection at surfaces, he had to endow the corpuscles with properties that made them rather like waves, a problem he did not resolve. A first glimpse of **wave–particle duality**!

The origin of the wave theory of light is usually attributed to Huygens on which he published a treatise in 1690. To explain the detection of light in the geometrical shadow when light is passed through an aperture, he proposed that light consists of a wavemotion, propagated through a continuous medium, the aether, with each point on an advancing wavefront envisaged as acting as a secondary source of wavelets (**Huygens' Principle**). The existence of such an 'aether' had already been envisaged by Newton as a hypothetical medium filling all space for the purpose of transmitting gravitational interactions, but not light. Strangely, Huygens envisaged the 'waves' not as continuous wavetrains but as a series of random pulses, and although the model explained refraction and reflection, it did not explain diffraction, interference, or polarisation. However, his theory was over-shadowed by the authority of Newton, even though earlier work by Grimaldi in 1660, and by Hooke in 1672, had already demonstrated blurring into geometrical shadows, consistent with Huygens' Principle.

Huygens did not need to specify the type of wavemotion he envisaged, but it was thought to be a longitudinal wavemotion, as with sound. It seemed that Young too may have had that view initially, but it was his double-aperture experiment in 1801 and the work of Fresnel around 1815 that really established the wave theory familiar today. Malus, Young, Arago, and especially Fresnel, all led the way to a model of light as a transverse vibration in the aether, and it was Maxwell's electromagnetic theory of 1864 that later suggested that it consists of oscillating electric and magnetic fields travelling through this (assumed) aether.

In Young's original historical experiment sunlight passing through a pinhole (Fig. 2.1) illuminated a screen some distance away, containing two closely-separated pinholes A and B. On another screen, D, about as far away again, fringes of light and dark were observed. Neither pinhole alone produced the fringes and their existence was interpreted as interference between light spreading as **'secondary wavelets'**

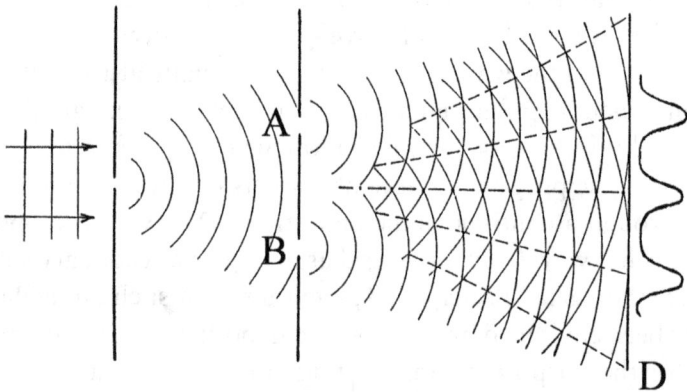

Figure 2.1. Young's experiment.

from the two pinholes. (This is often performed, and referred to, as a **'double-slit experiment'**.)

However, evidence for a particle model of light came with Einstein's 1905 interpretation of the photoelectric effect (see Appendix F) and he soon envisaged a need for a wave–particle duality theory. A striking example of the need for that was demonstrated in 1909 by G. I. Taylor. Using the same type of experimental set-up as in Young's experiment, but with a light source so weak that the image on the screen formed very slowly, a large number of spots were observed to accumulate and build up the bands observed straightaway with a brighter source. This provided visual evidence of the interference associated with wave motion, yet at the same time evidently resulting from the passage of 'particles' — the **'light quanta'** of Einstein in his explanation of the photoelectric effect, and further demonstrated by the **Compton Effect** of 1923.

The concept of wave–particle duality was extended to matter in general by de Broglie and others. Initially, it was demonstrated experimentally by the diffraction of electrons (Chap. 7) and subsequently by the diffraction of e.g. α-particles, neutrons, fullerene molecules, which contain sixty carbon atoms in a spherical configuration (Arndt *et al.*, 1999), and more recently by $C_{60} F_{48}$ molecules (Hackermüller *et al.*, 2003).

We return to the interpretation of these phenomena in later chapters. Meanwhile, we need to note the way in which the mystery of the 'aether' was resolved because this led to the birth of Einstein's **Special Theory of Relativity** in 1905, which played a crucial role in the development of the quantum theory. (Einstein's theory was so named to distinguish it from his 1915 **General Theory of Relativity** which dealt with gravity.)

2.2. The 'Aether'

2.2.1. *The stellar 'aberration of light'*

In 1726 James Bradley found that in his measurements of the distances of the stars — using their changing parallax against the background of more distant stars as the Earth travelled round its orbit — there was an effect that implied that the light from the stars was travelling through a medium, called the '**luminiferous aether**' at that time, that was not moving with the Earth. If the aether travelled with the Earth, then light from the star would travel straight down the axis of a telescope pointed towards the star. Instead, he found that the telescope had to be slightly deflected from that direction, rather like tilting an umbrella carried when walking through rain. From the deflection and the known speed of the Earth in its orbit, he obtained a value for the speed of light that was in agreement, within experimental limits, with the accepted value at that time (Rømer's method of 1675).

Bradley had assumed that light consisted of corpuscles, but when light was established as a transverse wavemotion in the early 1800s, as we have seen above, Young showed that stellar aberration could equally well be explained on the wave theory. Young believed that the aether "pervades the substance of all material bodies with little or no resistance, as freely perhaps as the wind passes through a grove of trees" (Whittaker — see Bibliography).

However, various questions arose and the mystery of the aether continued. For example, would light rays coming from a star be refracted differently from rays originating from terrestrial sources?

2.2.2. *Arago's experiments*

In 1810, Arago reported (and published much later) the results of experimental tests to ascertain, for example, whether the refraction of light from a star by a prism was influenced by the motion of the Earth through the aether. He found no such influence, and the obvious explanation appeared to be that the aether travels with the Earth. This work by Arago receives scant, if any, attention in most books, yet it was crucial at the time. For a good accessible account of this and the topics in the above section, see Whittaker (Bibliography). Fresnel set out to resolve the conflict between Arago's findings and the stellar aberration effect, as we shall see next.

2.2.3. *Fresnel's 'drag' mechanism (1818)*

Young, in his studies of **Newton's rings,** had concluded that the speed of light is reduced in dense media. Fresnel adopted this idea and proposed that the density of aether in any body is proportional to the square of its refractive index. Also, he assumed that when such a body is in motion, the excess part of the aether comprising the excess of its density over the density of aether in vacuo, is carried along with it, the rest of the aether remaining stationary. Thus, only the aether of empty space was 'stationary'.

Using this idea, Fresnel was able to account for both stellar aberration and Arago's findings concerning refraction.

Support for the '**drag hypothesis**' came with various experiments, initially those of Fizeau in 1851 who demonstrated the drag effect in the passage of light through a moving column of water. In 1871, Airy verified a prediction by Fresnel, viz. that if a telescope is filled with water, the observation of stellar aberration is unaffected. (At that time, other matters were resolved, such as Fresnel's clarification of polarisation as evidence that the wavemotion of light is a transverse vibration rather than longitudinal, as had been previously assumed on the basis of regarding it as similar to sound.)

2.2.4. *The 1887 Michelson–Morley experiment*

This historic experiment (and the earlier, less sensitive initial experiment in 1881 of the same type, by Michelson) attempted to test unequivocally whether or not the Earth is travelling through an 'aether'. The experiment (well documented in most textbooks) amounted to comparing the time taken for light to travel a given distance through the assumed stationary aether of space (i) in the same direction as the Earth was moving in its orbit, and (ii) in a direction at right angles to that. No difference was found despite the great precision of the final experiments. It could only mean that the aether was travelling with the Earth.

This was therefore at variance with Fresnel's 'drag hypothesis' and the Bradley observation of stellar aberration — both indicating the aether in space to be stationary.

An equally significant observation was that the speed of light in space is the same regardless of the motion of the source or the observer. Constancy of the speed of light was already embodied in **Maxwell's Electromagnetic Theory** of 1864, but realisation of its full implication awaited Einstein's **Special Theory of Relativity** in 1905. Meanwhile, other explanations were offered for the findings of the Michelson–Morley experiment.

Michelson was awarded the 1907 Nobel Prize for physics for his "optical instruments and investigations".

2.2.5. *The FitzGerald contraction*

In 1889, FitzGerald pointed out that the Michelson–Morley result would be explained if the linear dimensions of solids contract very slightly in the direction of their travel — the so-called '**FitzGerald contraction**'. This, and particularly the work of H. A. Lorentz who also contributed to the idea of contraction, proved to be a prevision of Einstein's Special Theory of Relativity. At this time, FitzGerald was also pondering the possibility of an upper limit to the speed of light.

Both FitzGerald and Lorentz (with varying knowledge of each other's work) were also active in investigating the aftermath of Maxwell's Electromagnetic Theory. It was Lorentz's contribution in particular that led to another approach to the contraction hypothesis.

2.2.6. *The Lorentz contraction*

From 1892, Lorentz explored (in several stages) the fact that Maxwell's equations only deal with how electromagnetic fields propagate for a given distribution of electric charges and currents. The equations did not deal with how the charged entities (e.g. 'electrons' as later named) themselves behave. Investigating this led Lorentz to derive his '**Equations of Motion**' and the '**Lorentz Force**', and he also proved that an accelerated electron must radiate energy.

With the problem of the aether still unresolved and much discussed, Lorentz also investigated the influence of the Earth's motion on Maxwell's Electromagnetic Theory. By 1895, he had shown that

Maxwell's equations for a moving system retain the same form as for a system at rest if time as a variable is modified by the speed of the system (and the speed of light) — what he called a 'local time'. This resulted in the famous 'Lorentz Transformation' which dealt with the dependence of both dimensions and mass on speed, significant with speeds approaching that of light. Note that this was based on his atomistic interpretation of the Maxwell equations in terms of charges and currents carried by particles (later called 'electrons').

Like the FitzGerald contraction hypothesis, the Lorentz Transformation provided (in a different way) a contraction explanation of the Michelson–Morley nil result. Hence, the often-used combined expression 'Lorentz–FitzGerald contraction' or vice versa.

The details of this work by Lorentz and others can be omitted here because various experiments showed the falsity of contraction hypotheses based on the ideas at that time. For example, if such a contraction of materials did occur, then an isotropic material such as glass might be expected to show double refraction (Lord Rayleigh, 1902). Rayleigh and Brace in 1904 with more refined experiments, failed to observe such an effect.

* * *

This summary indicates something of the state of knowledge on these problems when Einstein produced his Special Theory of Relativity in 1905.

2.3. Einstein's Special Theory of Relativity (1905)

Einstein's Special Theory of Relativity was to become intimately involved in the development of the quantum theory. Though a separate branch of physics, we need to see how the theory came about and to establish just those features that we need in the following chapters.

2.3.1. *Introduction: reference frame transformations*

In essence, Einstein's Special Theory of Relativity resolved the problem posed by the Michelson–Morley experiment as well as the constancy of the speed of light already implicit in Maxwell's Electromagnetic Theory.

The biography of Einstein by Abraham Pais (see Bibliography) shows the complexity of the interactions and the developments in the ideas of a number of people at the time leading up to Einstein's theory — notably Poincaré and Larmor. With hindsight, it is now recognised that had Lorentz appreciated the full significance of space and time as defined by the transformations he had devised, he would have been the author of the theory of relativity. However, Einstein developed his theory without reference to Lorentz's transformations, and in noting this, Pais says that Einstein 'invented them himself' in the context of '**two postulates**' based, as mentioned above, on the implications of the Michelson–Morley experiment and the constancy of the speed of light:

(i) The laws of physics need to be expressed in equations having the same form in all frames of reference moving at constant speed regardless of the value of the speed, i.e. '**inertial frames**'. (An inertial frame is a reference frame in which bodies move in straight lines with constant speeds unless acted on by external forces — thus it is a reference frame in which free bodies are not accelerated.) This means that one can never detect the absolute motion of bodies through space, only the motion of a body relative to another.

(ii) The speed of light is the same regardless of whether the light is emitted by a source at rest relative to the observer or in uniform motion relative to the observer.

Einstein regarded the confusing results of attempts to detect an aether as evidence that the speed of light is the same for all observers. Furthermore, the constancy of the speed of light as embodied in Maxwell's Equations was of particular significance to him, perhaps more so than the Michelson–Morley experiment which also pointed to this conclusion. In introducing his famous 1905 papers, he stated that with his two postulates, even the concept of an aether and of space at absolute rest (and endowed with special properties) would be superfluous.

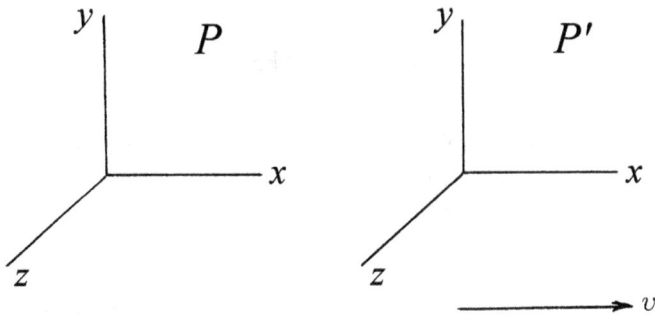

Figure 2.2. Frame P' moves in the $+x$-direction with speed v relative to frame P.

Einstein went into all this in enormous detail, turning to a re-examination of the basic problems of **kinematics** (which deals with the motions of bodies but not with the forces causing them — the province of **dynamics**). We can best appreciate the main result that we are going to be concerned with by considering the simple case of an event seen from two different frames of reference P and P' in Fig. 2.2. P' is moving at constant speed v in the $+x$-direction of P, and we can set the time at zero when they coincided. Now, suppose an observer in frame P observes an event at $x, 0, 0$ at time t (we can keep to one dimension for simplicity). By then, frame P' has moved in the x direction by distance vt, so relative to the frame P' the event has coordinates

$$\begin{aligned}
x' &= x - vt, \\
y' &= y, \\
z' &= z, \\
t' &= t.
\end{aligned} \qquad (2.1)$$

This is a classical **Galilean transformation**.

However, if c is the speed of light in frame P, then in frame P' it would be

$$c' = c - v.$$

To resolve such a violation, we start by introducing a factor γ into the equation for x':

$$x' = \gamma(x - vt), \qquad (2.2)$$

and we say that γ must not depend on x or t, but may be a function of v. (The other axes are not affected because they are normal to the direction of v.)

As the equations of physics must have the same form in both P and P', we can change the sign of v to write the corresponding equation for x in terms of x' and t':

$$x = \gamma(x' + vt'), \qquad (2.3)$$

where γ must be the same in both frames because there is no difference between P and P' other than the sign of v. The y and z coordinates are unaffected by the reversal of v because they are normal to v. However, $t \neq t'$ as we can see by substituting the value of x', given by Eqn. 2.2, into Eqn. 2.3:

$$x = \gamma[\gamma(x - vt) + vt'],$$

whence

$$t' = \gamma t + \left(\frac{1 - \gamma^2}{\gamma v}\right) x. \qquad (2.4)$$

The requirements of the first postulate are therefore met with the following transformations:

$$\begin{aligned}
x' &= \gamma(x - vt), \\
y' &= y, \\
z' &= z, \\
t' &= \gamma t + \left(\frac{1 - \gamma^2}{\gamma v}\right) x.
\end{aligned} \qquad (2.5)$$

(That there is no change in the y and z directions can easily be shown. See e.g. Bernstein *et al.*, 2000.)

To seek a value of γ that also allows the requirements of the second postulate to be met, imagine a light signal being sent off at the origins of P and P' when they coincided at $t = t' = 0$. An observer in each frame measures the speed of the light and although one frame is moving relative to the other, the measured speed must be the same in both frames.

Therefore in P:

$$x = ct,$$

and in P':

$$x' = ct'.$$

The second of these refers to the P' frame so we can substitute for x' and t' from Eqn. 2.5, which gives

$$\gamma(x - vt) = c\left[\gamma t + \left(\frac{1 - \gamma^2}{\gamma v}\right)x\right].$$

Solving for x gives

$$x = ct\left(\frac{1 + \dfrac{v}{c}}{1 - \left(\dfrac{1}{\gamma^2} - 1\right)\dfrac{c}{v}}\right),$$

but

$$x = ct,$$

whence

$$\gamma = \frac{1}{\sqrt{1 - v^2/c^2}}. \tag{2.6}$$

Substituting this expression for γ into Eqn. 2.5 gives

$$x' = \frac{x - vt}{\sqrt{1 - v^2/c^2}},$$
$$y' = y,$$
$$z' = z, \tag{2.7}$$
$$t' = \frac{t - vx/c^2}{\sqrt{1 - v^2/c^2}}.$$

Known as **Lorentz Transformations** (as applied here to v acting in just the x direction for simplicity), transformations of this type were originally derived in a different way by Lorentz in the context of an investigation of Maxwell's equations. Lorentz had regarded the transformations as a convenient mathematical tool for proving that terrestrial optical experiments are independent of the motion of the Earth. Thus, he had gone some way to explaining the absence of any evidence for a stationary **aether**.

For observations made in frame P', the corresponding equations are

$$x = \frac{x' + vt}{\sqrt{1 - v^2/c^2}},$$
$$y = y',$$
$$z = z', \tag{2.8}$$
$$t = \frac{t' + vx'/c^2}{\sqrt{1 - v^2/c^2}}.$$

These '**Inverse Lorentz Transformations**' are consistent with the contraction ideas that were arrived at earlier as a purely mechanistic explanation of the Michelson–Morley nil result. Suppose we have a rod in frame P of length l extending from x_1 to x_2. What is its length, $l' = x'_2 - x'_1$, as determined by an observer in frame P' travelling away from P in the x direction?

To an observer making measurements from P', the coordinates x'_1 and x'_2 of the ends of the rod in P are given by using Eqn. 2.8, viz.:

$$x_1 = \frac{x'_1 + vt}{\sqrt{1 - v^2/c^2}},$$

and

$$x_2 = \frac{x'_2 + vt}{\sqrt{1 - v^2/c^2}},$$

which gives

$$x'_2 - x'_1 = (x_2 - x_1)\sqrt{1 - v^2/c^2},$$

whence

$$l' = l\sqrt{1 - v^2/c^2}. \tag{2.9}$$

Thus, the rod as measured from P' has shortened.

A similar effect occurs with time — '**time dilation**'. An event taking a certain time in one frame will take a different time as measured by an observer of that event but travelling in another frame. The equation corresponding to Eqn. 2.9 is

$$t' = \frac{t}{\sqrt{1 - v^2/c^2}}. \tag{2.10}$$

Note that the Lorentz Transformations correctly reduce to the classical equations when the relative speed of the reference frames is small compared to the speed of light.

(The objective meaning of simultaneity was discussed by the great French mathematician Henri Poincaré from a different philosophical starting point in 1898 when he wrote that "we have no direct intuition about the equality of two time intervals", and in 1904 he envisaged — as FitzGerald already had in 1889 — that the speed of light "would become an impassable limit".)

We have seen that Einstein arrived at the Lorentz Transformations solely from the starting point of his two postulates. They made the notion of an aether and the concept of stationary space redundant. Each observer carries his own system of space and time, and the units he employs for measurement adjust themselves automatically to eliminate the influence of (unaccelerated) motion. However, it is neither practical nor necessary to reproduce here the very detailed studies of the consequences of the **Special Theory of Relativity**. It will suffice to note the relativistic aspects of energy and momentum which we shall need in later chapters. These are briefly dealt with in the following section.

2.3.2. *Energy and momentum*

Energy and momentum are important in physics because they are conserved quantities: the total energy and the total momentum of a system are each unchanged by interactions within the system.

With the establishment that there are no preferred observers (i.e. no universal basic reference frame) it is essential to arrange that all observers see the same conservation rules. The pre-relativity definitions of momentum and energy therefore had to be modified.

In the Special Theory of Relativity, energy is expressed (Einstein, 1905(d)) as

$$E = \frac{mc^2}{\sqrt{1 - v^2/c^2}}, \tag{2.11}$$

and momentum (Planck,1906(a)) as

$$p = \frac{mv}{\sqrt{1 - v^2/c^2}},$$
(2.12)

where v has the same meaning as in the previous section.

Eliminating v^2/c^2 gives

$$E^2 - (pc)^2 = (mc^2)^2,$$

or

$$E = \sqrt{p^2c^2 + (mc^2)^2},$$
(2.13)

commonly referred to as the **fundamental equation of the Special Theory of Relativity.**

This equation is independent of the observer and therefore meets the requirement of invariance with respect to inertial frames of reference (§2.3.1). Although two different observers will obtain different results, $E_1 \neq E_2$ and $p_1 \neq p_2$, both will agree with Eqn. 2.13 because

$$E_1^2 - (p_1c)^2 = E_2^2 - (p_2c)^2 = (mc^2)^2.$$
(2.14)

For a mass particle at rest relative to a particular observer, Eqn. 2.13 gives

$$E_0 = mc^2,$$
(2.15)

which is Einstein's well-known expression relating mass and energy for a particle at rest (i.e. momentum zero). The original concept of 'rest mass', m_0, regarding mass as varying with velocity, lost favour on the basis that mass as such is a fixed property of a body. It is the momentum that changes with velocity (Eqn. 2.12).

Einstein's original derivation, not used above, was given in his famous papers of 1905. Strangely, that work received little attention until nearly a year later when Planck drew attention to its profound importance. Plank contributed to the development of the theory, and was the first to introduce it into quantum theory, where for example in 1906 he showed that his **'quantum of action'**, h, is a relativistic invariant. Another important early development was made in 1908 by Minkowski who saw that space and time had to be regarded as a single entity, a **space–time continuum.**

*   *   *

Einstein's General Theory of Relativity, dealing with gravity, came ten years later and at that time he agreed (allegedly to please his revered father-figure, H. A. Lorentz) that the word 'aether' could be used to describe space–time, though without the properties previously ascribed to it (Kostro, 2001).

Chapter 3

Thermal Radiation and Planck's 'Energy Elements'

3.1. Introduction

By the end of the 19th century, Classical Physics had dealt satisfactorily with many aspects of physics but it had failed to account for the experimentally observed spectra from hot bodies and to properly explain experimental measurements of specific heats.

We will recall some details of these failures because only in understanding them can we appreciate not only the need for the concept of the quantum but also how it was introduced.

Classical Physics relied mainly on Newtonian Mechanics and Maxwell's Electromagnetic Theory. Little was known about the structure of matter, though the notion of the 'corpuscular' nature of matter goes back some centuries, and the idea of the atom became accepted in Newton's time when he wrote in 1718 of hard, impenetrable, particles.

By 1800, much progress had been made concerning the atomic nature of chemical elements (cf. Darwin's atom–molecule theory), but the discovery of the electron by J. J. Thomson in 1897 was the crucial first step made to probe the actual structure of the atom. He was awarded the 1906 Nobel Prize for physics for "theoretical and experimental investigations of the passage of electricity through gases".

However, Thomson's model was of electrons embedded in a positively-charged mass. Not until 1911 did we have Rutherford's planetary model of the nuclear atom (which resulted from his observation on the scattering, by metal foils, of α-particles from radioactive substances (spontaneous radioactivity having been discovered in 1896 by Becquerel — who shared the 1903 Nobel Prize with Pierre and Marie Curie). Then in the following year, 1912, the Laue Experiment and especially the work of W. H. and W. L. Bragg revealed for the first time the actual atomic structure of matter.

When the quantum entered physics in 1900, the macro-properties of materials were still having to be interpreted in terms of statistical distributions of the most probable arrangements at the micro level.

This statistical approach was largely the work of Boltzmann, Gibbs, and Maxwell. It had led, among other things, to a **Statistical Mechanics** interpretation of the **Second Law of Thermodynamics.**

As noted above, a failure of classical methods arose in connection with attempts to explain the observed spectra of thermal radiation, and experimental measurements of the specific heats of solids.

With regard to specific heats, according to the principle of the **Equipartition of Energy** if an atom has n degrees of freedom, the average energy of each atom is nkT where k is the **Boltzmann's Constant** and T is the absolute temperature. Thus, for a gm-atom the average energy E is given by

$$E = nNkT, \tag{3.1}$$

where N is Avogadro's number.

Then the specific heat, C_v, at constant volume, is given by

$$C_v = \frac{dE}{dT} = Nnk, \tag{3.2}$$

which is a constant and in good agreement with values experimentally determined at room temperature (cf. **Dulong and Petit's Law**) (Fig. 3.1). However, at lower temperatures, specific heat diminishes and the theory does not explain this. It was solved with the quantum theory by Einstein in 1907 (§4.5).

We turn now to the failure of classical methods concerning thermal radiation.

To provide a basis for theoretical work on thermal radiation, the concept of 'black-body radiation' (or 'cavity radiation') was widely

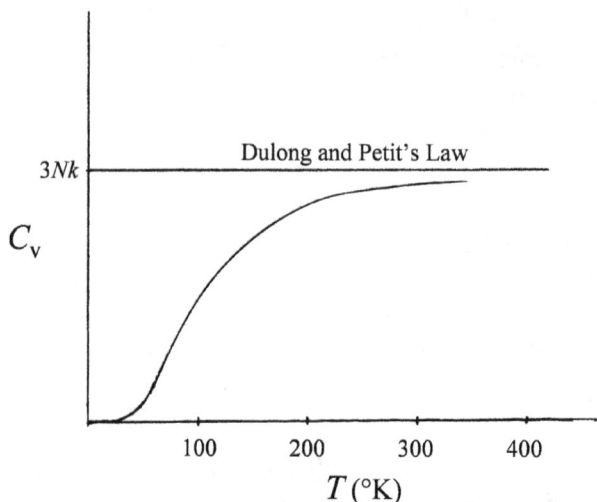

Figure 3.1. General form of the variation of the specific heat of metals with temperature, compared with Dulong and Petit's Law.

used during the 19th century. Kirchhoff had found that the ratio of the emissive and absorptive powers of a body depends only on the temperature of the body, not on its composition (**Kirchhoff's Law** of 1859). This led Kirchhoff to define a perfect 'black-body' as one which at all temperatures absorbs all the radiation falling upon it for all wavelengths (Chap. 1). Such black-body radiation is achieved with a cavity made of highly-insulating walls and closed except for a very small orifice. Held at a uniform temperature, the radiation from the orifice is a representative sample of the equilibrium radiation inside and is independent of the materials of the wall, depending only on their temperature. An equilibrium spectral distribution independent of everything except the temperature was clearly of great fundamental importance. It was in investigating this that the difficulties arose.

The distribution of energy in black-body radiation with respect to frequency/wavelength had been investigated experimentally, notably by Lummer and Pringsheim in 1897–1899. The general form of their results is shown in Fig. 3.2. Such results fitted Stefan's empirical law of 1879 (based on measurements by Tyndall, of heat loss from a hot platinum wire), which stated that the total heat loss (which would

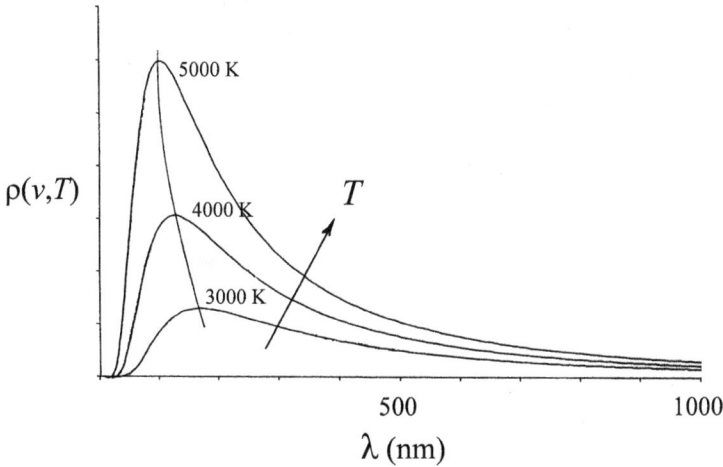

Figure 3.2. Wien's Displacement Law (Eqn. 3.6).

correspond to the area under curves such as in the above figure) for an emitter at the absolute temperature T is given by

$$E \propto T^4. \quad \text{Stefan's Law} \qquad (3.3)$$

A theoretical derivation of **Stefan's '4th power law'** was given by Boltzmann in 1884 (see e.g. Whittaker — see Bibliography) by applying the Second Law of Thermodynamics to **radiation treated as a gas** whose pressure was regarded as the **radiation pressure** in Maxwell's Electromagnetic Theory. The '**Stefan–Boltzmann Law**' is written as

$$E = \int_0^\infty \rho(v, T)dv = \sigma T^4, \qquad (3.4)$$

where E is the total energy density, and $\rho(v, T)$ is the '**energy distribution function**' — the energy per unit volume at frequency v in a cavity at absolute temperature T. σ is known as the '**Stefan–Boltzmann Constant**'.

An expression for $\rho(v, T)$ was given by Wien in 1894. Using Stefan's empirical T^4 law and reasoning about the conditions required for the radiant energy distribution to be in equilibrium with an enclosure at temperature T, he deduced, albeit with little theoretical justification, the following:

$$\rho(\nu, T) = \nu^3 f(\nu/T),$$

or in terms of wavelength,   (3.5)

$$\rho'(\lambda, T) = \lambda^{-5} f'(\lambda, T).$$

(If $d\lambda$ corresponds to $d\nu$, then $\rho' d\lambda = \rho d\nu$, and with $\nu = c/\lambda$, this gives the first expression, f' and f being different functions.)

[Note that Eqn. 3.5 is in accord with Stefan's law,

$$E = \int_0^\infty \rho(\nu, T) d\nu = \int_0^\infty \nu^3 f\left(\frac{\nu}{T}\right) d\nu = T^4 \int_0^\infty x^3 f(x) dx,$$

where $x = \nu/T$ is a new variable of integration, and the last integral is a pure number.]

Equation 3.5 is known as **Wien's Theorem**, or **Wien's Displacement Law** because the peaks of the curves for different temperatures shift towards higher frequencies with increasing temperature (Fig. 3.2). It easily follows from Eqn. 3.5 that

$$\lambda_{max} T = \text{constant},$$ (3.6)

where λ_{max} is the peak wavelength. It is this equation, not Eqn. 3.5, that is often referred to as **Wien's Displacement Law**. (It can of course easily be derived from Eqn. 3.5.) Wien was awarded the 1911 Nobel Prize for physics for his "discoveries regarding laws governing the radiation of heat".

Experimental confirmation of Eqns. 3.5 and 3.6 was obtained by various workers, notably by Paschen, Lummer and Pringsheim.

Because of the profoundly fundamental nature of black-body radiation (its independence of the chemical composition, properties, shape and size of the materials of the source), the elucidation of the function $f(\nu/T)$ was of great importance.

The close similarity between the form of the black-body radiation spectrum and the **Maxwell distribution of velocities** (Appendix B) led to the postulate that Maxwell's formula also holds for the 'molecules' of the radiating body.

Assuming that the vibrations of a molecule produce radiation with frequency and intensity dependent only on its velocity, and applying Maxwell's distribution law, led Wien in 1896 to propose the exponential form for $f(v/T)$:

$$f\left(\frac{v}{T}\right) = \alpha e^{-\beta v/kT}, \tag{3.7}$$

with α and β as constants. [Maxwell's distribution of velocities (see Appendix B) gives the number of molecules with velocity v as

$$Cv^2 e^{-mv^2/2kT}.$$

Wien was taking the intensity of the radiation as proportional to the number of molecules, and its frequency v as a function of molecular velocity.]

Equation 3.5 then becomes

$$\rho(v, T) = \alpha v^3 e^{-\beta v/kT},$$

or

$$\rho'(v, T) = C_1 \lambda^{-5} e^{-C_2/\lambda T}. \tag{3.8}$$

Known as 'Wien's Distribution (or Radiation) Law', the equation fitted the limited experimental black-body spectra at that time (Fig. 3.3). However, if $T = \infty$, it wrongly gives $\rho(v, T)$ which is still finite. It also fails for long wavelengths — increasingly so as temperature increases.

This was the situation when Planck turned his attention to the problem of finding a scientific basis for a detailed description of black-body radiation. Wien's distribution law was to be a very important stepping stone in his studies, and also to Einstein later, who regarded it as significantly and importantly correct, albeit as a limiting case for large v/T.

As we have already seen, drawing on a parallel between equilibrium radiation in a black-body cavity and the equilibrium behaviour of gases in an enclosure has permeated much of this work and it has continued to do so.

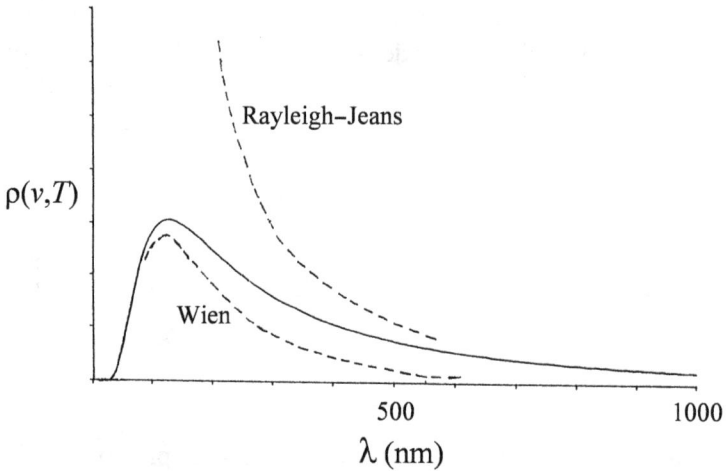

Figure 3.3. Observed black-body radiation distribution showing the misfit of Wien's Distribution Law (Eqn. 3.8) at long wavelengths and the failure of the Rayleigh–Jeans Law (Appendix D) at short wavelengths.

3.2. Planck and 'Energy Elements'

3.2.1. *Introduction: background*

In the years leading up to his successful explanation in 1900 of the black-body radiation spectrum, Planck's career was greatly influenced by the work of Clausius on the interpretation of the laws of thermodynamics. He was especially interested in the concept of irreversibility in the context of the Second Law — a law that, to paraphrase Clausius' interpretation, states that the entropy of an isolated system always increases, or remains constant (see Appendix A). He could not accept Boltzmann's reformulation of the Second Law as a statistical law, in which increase in entropy would only be highly probable rather than an absolute certainty.

Planck's interest in the thermodynamics of thermal radiation was probably aroused by Wien's 1894 paper on the Displacement Law (Eqn. 3.5), and from 1895 to 1900 he was engaged in detailed investigations. His publications (Planck, 1958) were profuse, and in the following notes we can outline the resulting features that were directly involved in arriving at the final solution of the black-body radiation problem. Planck later described this period as "the long and multiply-twisted path" that led to the quantum theory. Of the many authors who have written in great detail, and some at great length, about Planck's work in this area, the main ones include (see Bibliography) Gearhart, Jammer, Kangro, Klein, Kuhn and Pais. Extensive bibliographies are included by Kuhn and Kangro.

Planck started out on this work in 1895 and published two papers (Planck, 1896–1897) on an analysis of the behaviour, in an electromagnetic field, of a harmonic oscillator in the form of a **linear oscillating electric dipole** (the simplest body that can absorb and emit radiation), which is commonly referred to as a '**resonator**'. He had been impressed by Hertz's discovery in 1887 of the electromagnetic waves emitted by oscillating charges, and whose calculations, published in 1889, he used as his model. Planck followed this with a five-part series of papers (Planck, 1897–1899) on further aspects of the subject, entitled (in translation) "On Irreversible Radiation Processes". He introduced the idea of a resonator losing energy by conservative (i.e. non-dissipative) 'damping' to an electromagnetic field. By this means, he aimed to show (using only Maxwell's Laws) that the interaction of a resonator as it came into equilibrium with an electromagnetic field was irreversible, and was associated with what could be regarded as **electromagnetic entropy**. We may note here that Planck was encouraged in the validity of his model because of Kirchhoff's Law (§3.1), which stated that the black-body radiation spectrum was independent of the composition of the cavity walls.

However, Boltzmann pointed out that because Maxwell's Laws are invariant under time reversal, Planck had not demonstrated irreversibility. Forced to admit this, Planck introduced the concept of '**natural radiation**' (something akin to what we would call incoherent radiation) as an electromagnetic analogue of 'molecular disorder' as conceived earlier by Boltzmann.

He embarked on considering in more detail the nature of the model he was using. This resulted in a long paper under the same title as the five-part series, submitted in November 1899 (Planck, 1900), in which he noted that since his resonators are damped, they respond not to a single frequency but to a narrow range of frequencies. The latter could be described by a Fourier Series, and it was assumed that the Fourier components are independent. (This was the electromagnetic analogy to Boltzmann's concept of molecular disorder.)

In the course of these studies, he produced the 'Theorem' (Eqn. 3.9 in §3.2.2) that formed not only his starting point for tackling the black-body radiation problem but which was also to continue to play an important role in the development of the subject at that time.

Thus, we see that Planck was regarding black-body radiation as originating in (sub-microscopic) resonators in the walls of a 'vacuous' black-body cavity. The observed continuous spectrum of the radiation would be a manifestation of the enormous number of resonators with different frequencies. In the following sections, we distil the main, crucial steps that Planck took to achieve his goal. (Note that Planck was *not* assuming that wall materials actually consist of such resonators.)

There were essentially two stages. In the first, Planck arrived at an *empirical* energy-distribution function of black-body radiation that fitted well the experimental data at the time but which was theoretically unsatisfactory in that it was partly based on **Wien's Distribution Law** (Eqn. 3.8). In the second, Planck found a more satisfactory basis for his empirical equation, which necessitated the concept of '**energy elements**', named '**energy quanta**' in 1905 by Einstein (§4.2). However, as we shall see in the next chapter, it was Einstein who established the 'quantum' on a broader basis in one of his historic papers of 1905.

3.2.2. *Stage 1*

Planck's starting point (used again in Stage 2) was the equation, referred to in the previous section, that he had derived in the course of his studies. The equation related the black-body spectral distribution, $\rho(v, T)$, to the time-averaged equilibrium energy, $E(v, T)$, of a damped harmonic oscillator of frequency v at temperature T in the cavity wall. He had derived this by considering the equilibrium between the emission and absorption rates of radiation by damped resonators and obtained the equation,

$$\rho(v, T) = \frac{8\pi v^2}{c^3} E(v, T).$$
(3.9)

Here, the frequency v refers to both the radiation frequency and resonator frequency.

This equation, known as '**Planck's Theorem**', is regarded as one of his major contributions to Classical Physics and it played an important part in later developments including Einstein's subsequent studies. It first appeared in the final part of Planck's five-part publication of 1897–1899. Summary derivations are given by Kuhn (1978), Jammer (1989), and Pais (1982).

To complete the task of obtaining the radiation distribution $\rho(v, T)$, it was necessary to determine $E(v, T)$ in Eqn. 3.9, i.e. the average energy of a harmonic oscillator at temperature T. Although the **Equipartition Principle** could have provided an answer to this, Planck followed a thermodynamics route, seeking a relationship between energy and entropy, rather than one between energy and temperature. Fortunately, Planck did not use the equipartition principle (of which he may not have been aware) as it would have given him the wrong answer.

We do not know all the thoughts and ideas that Planck had at this particular time and there continues to be much speculation. In his numerous publications, he was certainly considering the problem in different ways. However, we can identify the key features that led to the establishing of an analytical expression that was in accord with experimental data for the black-body radiation spectrum, and also in accord with the Second Law of Thermodynamics.

To involve entropy as he wished, Planck used the expression from thermodynamics for a system at constant volume (see Appendix A):

$$\frac{dS}{dE} = \frac{1}{T}.$$
(3.10)

He went to the second derivative, however, because he said that it is the change in entropy increase that has physical meaning, unlike entropy itself. This gave

$$\frac{d^2S}{dE^2} = -\left(T^2\frac{dE}{dT}\right)^{-1}. \tag{3.11}$$

To evaluate this, Planck needed dE/dT. At the time, Wien's distribution law (Eqn. 3.8) fitted the available experimental black-body spectral data well. Substituting that into Planck's equation (Eqn. 3.9) gave

$$\alpha v^3 e^{-\beta v/kT} = \frac{8\pi v^2}{c^3}E(v, T),$$

i.e.

$$E(v, T) = \frac{\alpha v c^3}{8\pi}e^{-\beta v/kT}, \tag{3.12}$$

though of course there were still unknowns.

Using Eqn. 3.12 to obtain dE/dT gave

$$\frac{dE}{dT} = \frac{\beta v E}{T^2}$$

and Eqn. 3.11 then became

$$\frac{d^2S}{dE^2} = -\frac{1}{CE} \tag{3.13}$$

where C is a constant dependent on the resonance frequency but not temperature.

Planck was apparently intrigued by the simplicity of Eqn. 3.13 but he recognised that Wien's Distribution Law did not have a sound theoretical basis even though it fitted all the experimental data at that time. However, more recent work by Lummer and Pringsheim, reported in 1899, was revealing a misfit at low temperatures. Equation 3.13 was therefore not totally correct, though it merited respect. Then, on 7 October 1900, Rubens visited Planck and reported that he and Kurlbaum had just found, by extensive and very precise work, that at high temperatures and low frequencies, $\rho(v, T)$ was actually proportional to T. This certainly was not in Wien's distribution law (Eqn. 3.8).

Immediately after Rubens left, Planck worked on an interpolation that successfully combined Wien's distribution law with the need for linearity with T at small v/T.

[At that time, Rayleigh's theoretical investigation of black-body radiation was also indicating a linearity with temperature at low frequencies (see Appendix D), but Planck was allegedly unaware of, or indifferent to Rayleigh's work (which in any case had a major flaw in other aspects). It was the Ruben's and Kurlbaum's data that were significantly important to him. Rayleigh's method, which used the Equipartition Principle, gave a result that failed to match experimental data at high frequencies (Fig. 3.3), the so-called **UV catastrophe**. Fortunately, as mentioned earlier, Planck was not following that route though it is not certain why, perhaps because he was unaware of it or because it had been controversial (see Jammer, 1989 — see Bibliography).]

With $\rho \propto T$ at small v/T, as Rubens had just reported, Planck's initial distribution function (Eqn. 3.9) meant that

$$E(T) \propto T, \tag{3.14}$$

and therefore,

$$\frac{dE}{dT} = C_1$$

where C_1 is a constant. Using $dS/dE = 1/T$ again gave

$$\frac{d^2S}{dE^2} = -\frac{1}{T^2 C_1}$$

and with $E(T) \propto T$, this could be written as

$$\frac{d^2S}{dE^2} = -\frac{1}{C'E^2} \tag{3.15}$$

where C' is another temperature-independent constant.

Hence Planck had Eqn. 3.13 applicable over most of the frequency range, except for low v/T, where Eqn. 3.15 applied.

He found that combining the two in the following way:

$$\frac{d^2S}{dE^2} = -\frac{1}{E(a+E)}, \tag{3.16}$$

where a is a temperature-independent constant (though perhaps dependent on frequency), gave a distribution that covered the whole range well. He called this his '**entropy equation**'. To obtain the distribution, we proceed as follows.

Integrating Eqn. 3.16 gives[a]

$$\frac{dS}{dE} = \frac{1}{a} \log_e \left(\frac{a+E}{E} \right) + b$$

$$= \frac{1}{T} \quad \text{by Eqn. 3.10,}$$

where b is a constant.

Thus we have

$$\frac{1}{a} \log_e \left(\frac{a+E}{E} \right) + b = \frac{1}{T}.$$

The Rubens and Kurlbaum linearity result required that both sides of this equation must approach zero at high temperatures (and therefore high energies), so $b = 0$. The above equation could therefore now be written (using $\log_e x = a, x = e^a$) as

$$E = \frac{a}{e^{a/T} - 1}. \tag{3.17}$$

Substituting this into Planck's initial distribution function (Eqn. 3.9) gives

$$\rho(v, T) = \frac{8\pi}{c^3} \left(\frac{av^2}{e^{a/T} - 1} \right).$$

To have v^3 in the numerator for Wien's Law (Eqn. 3.5), we can put $a = Bv$ with B a constant, giving

$$\rho(v, T) = \frac{Av^3}{e^{Bv/T} - 1}, \tag{3.18}$$

where A is a constant independent of v and T.

This was in good agreement with all the available spectral data, including the linearity observed by Rubens and Kurlbaum, as can be

---

[a] Use the standard integral $\int \frac{dx}{x(a+x)} = -\frac{1}{a} \log_e \left(\frac{a+x}{x} \right) + b.$

seen by noting that for very small v/T, Eqn. 3.18 reduces (using $e^x \approx 1 + x$ for small x) to

$$\rho(v, T) \approx \frac{A}{B} v^2 T. \tag{3.19}$$

Planck announced his result on 19 October 1900, barely two weeks after hearing of the Rubens and Kurlbaum data. He realised that his 'entropy equation' (Eqn. 3.16) was an inspired guess. He also realised that the Wien equation he had used was empirical but believed that the result he had obtained was correct — and indeed it was — so he set out to seek the correct theory behind it — and succeeded.

Before moving on to the major step ahead we should note that Planck's achievement caused Max Born (who became the 1954 Nobel Prize winner for his own contribution to quantum physics) to express the view that Planck's idea of combining the two relationships "was one of the most fateful and significant interpolations ever made in the history of physics; it reveals an almost uncanny intuition".

With the scene set, we can now appreciate the final development that led to the equation that introduced the quantum.

3.2.3. *Stage 2: Planck's final distribution function*

To derive a theoretical and physical basis for his new energy distribution function (Eqn. 3.18), Planck turned reluctantly from solely using classical thermodynamics, which could take it no further, to Boltzmann's statistical mechanics approach to the probabilistic aspects of thermodynamic equilibrium. Boltzmann had shown that, in the case of a gas in which the molecules do not conform to the Maxwell–Boltzmann distribution law, there is a readjustment to a configuration that approaches that distribution. In his 'H-theorem' he gave an expression for the rate at which this approach takes place. It involved a function, H, that decreases as equilibrium in accord with the Maxwell–Boltzmann distribution is reached.

Planck explored the analogy between this and the well-known increase in entropy in the Clausius sense (see Appendix A) as equilibrium is approached. However, he was convinced that entropy is not inherently probabilistic and he set about reinterpreting

Boltzmann's probabilistic theory in his own way. Jammer (1989 — see Bibliography) had summarised this by saying that in Planck's '**combinatorial approach**', he was merely determining the "total number of possible complexions and not, as for Boltzmann, the number of possible complexions corresponding to the macro-state which can be realised by the largest number of complexions". ('**Complexion**' is a term allegedly introduced by Boltzmann.)

Planck said he "simply postulated" that

$$S = k \log W, \tag{3.20}$$

where W is the 'probability' (i.e. Planck's calculation of the number of possible complexions) of the state of a system whose entropy is S (see Appendix A, §A.3).

To obtain an expression for $E(v, T)$ in Eqn. 3.9, Planck's intention now was (i) to use the above equation to determine the average entropy S of an equilibrium system of resonators of frequency v in terms of its probability W calculated as the number of ways that indistinguishable units of energy of that frequency can be distributed among distinguishable resonators, (ii) to use this entropy S to derive the average energy E, again using $dS/dE = 1/T$ (Eqn. 3.10). (The method of calculating W was to lead to much relevant discussions in the later developments.)

It was a wild stroke of genius, as we shall see. Now we turn to the main details.

With N resonators (frequency v, average energy E, and average entropy S) in the walls of a black-body cavity, the total energy, E_N, and total entropy, S_N, of the system are

$$\begin{aligned} E_N &= NE, \\ S_N &= NS. \end{aligned} \tag{3.21}$$

Equation 3.20 for the entropy of the system is then

$$S_N = k \log W, \tag{3.22}$$

and W is now the number of ways in which the total energy E_N can be distributed among the resonators according to Planck's method of calculation.

Planck assumed that the total energy E_N was made up of n small finite indivisible units each of energy magnitude ε, so we have

$$E_N = n\varepsilon. \tag{3.23}$$

The number of ways, W, that the n *indistinguishable* units of energy can be distributed between N *distinguishable* resonators is given by standard combinatorial analysis as

$$W = \frac{(N + n - 1)!}{n!(N - 1)!}. \tag{3.24}$$

(Note: This was not a conventional **Boltzmann Statistics** method, which deals with the distribution of *distinguishable* entities (see Appendix B, §B.2.)

As N and n are very large, one can write

$$W = \frac{(N + n)!}{n!N!}. \tag{3.25}$$

Applying Sterling's formula for large numbers,

$$\log x! \approx x \log x - x,$$

gives

$$S_N(= k \log W) = Nk \left[\left(1 + \frac{n}{N}\right) \log \left(1 + \frac{n}{N}\right) - \left(\frac{n}{N}\right) \log \left(\frac{n}{N}\right)\right]. \tag{3.26}$$

To introduce energy into this, we note that Eqns. 3.21 and 3.23 give

$$\frac{n}{N} = \frac{E}{\varepsilon}.$$

Substituting this into the above equation, we obtain

$$S_N = Nk \left[\left(1 + \frac{E}{\varepsilon}\right) \log \left(1 + \frac{E}{\varepsilon}\right) - \left(\frac{E}{\varepsilon}\right) \log \left(\frac{E}{\varepsilon}\right)\right]. \tag{3.27}$$

This represents the total entropy, which is N times the average entropy of the resonators (Eqn. 3.21). For one resonator, the average entropy is therefore

$$S = k \left[\left(1 + \frac{E}{\varepsilon}\right) \log \left(1 + \frac{E}{\varepsilon}\right) - \left(\frac{E}{\varepsilon}\right) \log \left(\frac{E}{\varepsilon}\right)\right]. \tag{3.28}$$

Then, using Eqn. 3.10,

$$\frac{dS}{dE} = \frac{1}{T},$$

to relate entropy with energy gives [using $d(\log x)/dx = 1/x$],

$$\frac{dS}{dE} = \frac{1}{T} = \frac{k}{\varepsilon}\left[\log\left(1 + \frac{E}{\varepsilon}\right) - \log\left(\frac{E}{\varepsilon}\right)\right], \qquad (3.29)$$

whence (using $\log e^x = x$), we have

$$E(v, T) = \frac{\varepsilon}{e^{\varepsilon/kT} - 1}. \qquad (3.30)$$

The final step is seen by comparing this with

$$E(v, T) = \left(\frac{c^3}{8\pi}\right) v f(v/T), \qquad (3.31)$$

which is obtained by substituting Planck's initial distribution function (Eqn. 3.9) into Wien's Displacement Law (Eqn. 3.5).

For Eqns. 3.30 and 3.31 to be compatible, the energy units, ε, need to be proportional to v, i.e.

$$E = hv, \qquad (3.32)$$

where h is a constant of proportionality.

Equation 3.30 for the **average equilibrium resonance energy** then becomes

$$E(v, T) = \frac{hv}{e^{hv/kT} - 1}, \qquad (3.33)$$

and Planck's expression for $\rho(v, T)$ in Eqn. 3.9 becomes

$$\rho(v, T) = \frac{8\pi v^2}{c^3} \frac{hv}{e^{hv/kT} - 1}, \qquad (3.34)$$

known variously as Planck's **radiation law, the radiation distribution function,** and **the radiation formula.**

This has the exact form of Planck's function from a few months earlier (Eqn. 3.18) but which, as we saw, had to be obtained by judicious interpolation to fit experimental data, including the new data of Rubens and Kurlbaum. In this new equation, it remained necessary for hv to be small for linearity at small v/T but large enough for hv/kT to be appreciable over the higher v/T range. Thus we have the following:

(i) For very small $h\nu/kT$, Eqn. 3.34 becomes

$$\lim_{h\nu \to 0} \rho(\nu, T) = \frac{8\pi\nu^2}{c^3}kT, \qquad (3.35)$$

(using $e^x \approx 1 + x$ for small x) which is Rayleigh's expression (see Appendix D), correct only in giving the linearity with T as found for very small ν/T.

(ii) For large ν/T, Eqn. 3.34 changes to the form,

$$\rho(\nu, T) = \frac{8\pi h\nu^3}{c^3}e^{-h\nu/kT}, \qquad (3.36)$$

which is Wien's law (Eqn. 3.8) but includes h (where $\alpha = 8\pi h/c^3, \beta = h/k$), which fitted the data well except for small ν/T.

Planck's innovation meant that the simple harmonic resonators in the black-body cavity walls emit and absorb energy discontinuously in multiples of small indivisible units that he called '**energy elements**'. We have

$$E = nh\nu, \qquad (3.37)$$

where $n = 0, 1, 2, \ldots$ and h is now referred to as **Planck's constant**. Initially, Planck did not consider E as an integral multiple of $h\nu$. He visualised it more as a mathematical device than as having any physical significance (Jammer, 1989).

As mentioned earlier, the expression '**energy quanta**' for $h\nu$ was coined by Einstein in 1905, and Eqn. 3.37 then became known as '**Planck's quantum condition**'. [Planck later called h the '**quantum of action**' because '**action**', as used in mechanics (as we note again in §7.3.5) has units of energy × time, as does h.]

Planck presented his result to the German Physical Society on 14 December 1900.

Much is made of the way in which he obtained the expression for W in Eqn. 3.24, and it certainly led to many discussions later. However, even if he had used conventional Boltzmann statistics in this particular instance, he would have arrived at the same expression (Eqn. 3.33) for average energy (see §3.2.4). Indeed, it was suggested by Rosenfeld (quoted by Jammer) that in choosing his method, Planck had been influenced by realising that in order that (using our nomenclature here) S_N in Eqn. 3.27 would fit his conjectural Eqn. 3.22, then W in the latter would have to be in the form shown in Eqn. 3.25. In that sense, one could say he was working backwards to find the statistics that would give the right answer. There is confirmation of this in an article Planck published in 1943.

Referring to Planck's introduction of **'energy elements'** and the unusual statistics, Pais (1982) had said "...the justification for Planck's two desperate acts was that they gave him what he wanted. His reasoning was mad, but his madness has that divine quality that only the greatest transitional figures can bring to science. It cast Planck, conservative by nature, into the rôle of a reluctant revolutionary."

Planck was aware of having used a device to get the right answer. Although there is evidence that he felt that his result had some profound implications, the philosopher and historian of science, Thomas Kuhn, believed that Planck did not initially appreciate the significance of his 'energy elements'. Planck was simply postulating that radiation is emitted in discrete units. Not interested in the details of the abstract oscillators/resonators, he wanted to keep his work firmly based on classical physics in which radiation is actually propagated as a wave motion.

As another historian of science, Helge Krath, had put it, "Like Copernicus, Planck became a revolutionary against his will". It was ten years or so before he even finally accepted the statistical nature of the Second Law of Thermodynamics.

Though Planck can justifiably be described as the father of the quantum — he was awarded the 1918 Nobel Prize in physics — the discovery of a fuller meaning of the quantum awaited the genius of Einstein to make the next decisive steps.

Planck was able to obtain values for h and k from the black-body radiation data. They were near to today's values which in SI units are

$$h = 6.63 \times 10^{-34} \text{Js},$$
$$k = 1.38 \times 10^{-23} \text{JK}^{-1}.$$

*   *   *

Planck continued his studies in this area especially in the problem of absolute values for entropy, and in 1906 introduced into his work the Willard Gibbs concept of '**phase space**' (see Appendix C). The concept was used a great deal after that by others in developing various aspects of quantum theory, and we refer to it again especially in the context of the background to Schrödinger's wave mechanics (Chap. 7). Notes on phase space can be found in Appendix C since we use it in various parts of this book.

In 1906, Planck was also the first to use the **Special Theory of Relativity** in quantum theory when he showed that his '**quantum of action**' is a relativistic invariant.

3.2.4. *Summary of the main features of Planck's final derivation of the distribution function*

Planck had used classical methods to derive the spectral distribution function $\rho(v, T)$ of black-body radiation in terms of the average equilibrium energy $E(v, T)$ of a simple harmonic resonator visualised as comprising the cavity walls,

$$\rho(v, T) = \frac{8\pi v^2}{c^3} E(v, T). \qquad [3.9]$$

To obtain $E(v, T)$, he used the thermodynamic relation,

$$\frac{dS}{dE} = \frac{1}{T}, \qquad [3.10]$$

and to calculate S, he postulated

$$S = k \log W, \qquad [3.20]$$

where k was a constant, and W the number of 'complexions' of a system whose entropy is S.

Assuming the energy to be made up of small, finite, indivisible, units of magnitude ε gave

$$E(v, T) = \frac{\varepsilon}{e^{\varepsilon/kT} - 1}. \qquad [3.30]$$

To find ε, Planck compared Eqn. 3.30 with

$$E(v, T) = \left(\frac{c^3}{8\pi}\right) v f(v/T), \qquad [3.31]$$

which is obtained by substituting Planck's original Eqn. 3.9 into Wien's law (Eqn. 3.5). The latter had been based on Stefan's empirical T^4 law (subsequently obtained theoretically by Boltzmann by applying the second law of thermodynamics to **radiation treated as a gas** — a recurring feature in so much of the development of quantum mechanics), together with reasoning about the thermal equilibrium between radiation and its enclosure.

For Eqn. 3.30 to be compatible with Eqn. 3.31, ε needed to be proportional to frequency, so Planck put

$$E = hv, \qquad [3.32]$$

with h a constant, giving Eqn. 3.30, the **average equilibrium energy of a harmonic oscillator** as

$$E(v, T) = \frac{hv}{e^{hv/kT} - 1}. \qquad [3.33]$$

Thus, Planck obtained his final radiation distribution formula,

$$\rho(v, T) = \frac{8\pi v^2}{c^3} \frac{hv}{e^{hv/kT} - 1}. \qquad [3.34]$$

In assuming energy to exist in small units in arriving at his W for $S = k \log W$, Planck was initially (but only initially) following the normal procedure of classical statistical mechanics. There, with energy having all possible values, one would then allow ε to become infinitesimal in magnitude and infinite in number. The average number of oscillators with energy, ε_1 say, would be given by

$$Ce^{-\varepsilon_1/kT},$$

where C is a constant, and the overall average energy of an oscillator would be

$$\frac{\displaystyle\int_0^\infty \varepsilon_1 e^{-\varepsilon_1/kT} d\varepsilon}{\displaystyle\int_0^\infty e^{-\varepsilon_1/kT} d\varepsilon},$$

where each energy is weighted by the Boltzmann factor, and where the denominator is for normalisation (see Appendix B). This would have given the average energy as the classical kT, as in Rayleigh's method (see Appendix D).

Note that Planck's own method, with $S = k \log W$, would have given the same classical result if he had not imposed quantisation. The expression (Eqn. 3.33) for the average energy would be

$$\lim_{h\nu \to 0} \left(\frac{h\nu}{e^{h\nu/kT} - 1} \right) = kT.$$

Also note that Planck's expression for the average resonator-energy would also have resulted from inserting quantisation into Boltzmann's statistics. Equation 3.38 would become

$$E(\nu, T) = \frac{\sum_{n} n h \nu e^{-nh\nu/kT}}{\sum_{n} e^{-nh\nu/kT}}$$

$$= \frac{h\nu}{e^{h\nu/kT} - 1},$$

which was the method Debye used retrospectively in examining Planck's radiation formula after Planck introduced his energy elements $h\nu$ (see Appendix E).

Chapter 4

Einstein and the Quantum

4.1. Introduction

In the years 1902–1925, Einstein's work on the development of quantum physics was prodigious and generally regarded as unequalled in the history of science. The following list gives an indication of the range of some of his many achievements during that period. After that time, and till the end of his life, he concentrated on a search for a unification of General Relativity with electromagnetism — a search continuing today to exercise the minds of others.

  (i) The extension of Planck's concept of discontinuous 'energy elements' in the generation of black-body radiation to show, quite independently, that the actual propagation of light has also to be thought of (contrary to Planck's belief) as taking place by means of discrete energy units — what Einstein called '**energy quanta**' (also '**light quanta**') [named '**photons**' by G. N. Lewis in 1926 by comparison with 'electron' (Stoney, 1894) and 'proton' (Rutherford, 1920)].

 (ii) An explanation of the photoelectric effect.

(iii) Advances in the study of Brownian Motion and topics associated with it.

(iv) The Special and General Theories of Relativity.

 (v) A quantum theory of specific heats, which saw the birth of modern Solid State Theory.

 (vi) Independent derivations of Planck's radiation formula.
 (vii) The association of momentum with light quanta.
 (viii) A wave–particle duality concept of electromagnetic radiation.
  (ix) Statistical aspects of transitions relating to both Bohr's theory of the atom and Planck's law.
  (x) Study of energy fluctuations in radiation, suggesting, independently, wave–particle duality.

Einstein's work on these topics often overlapped in its sequence of publication, and similar ideas were developed and applied in different ways in the many papers he published. His work was based essentially on statistics and thermodynamics, especially the rôle of energy fluctuations — the random instantaneous deviations that occur about an equilibrium mean. The latter gave him an interest in the importance of the **volume dependence of entropy.**

A brief chronological summary will give a sense of how these topics permeated the development of the subject, and we will then look at the main underlying principles and how they were applied in the various ways so effectively.

4.2. Overview

1902–1904

In three papers, Einstein laid down his own foundations of statistical physics. He had been preoccupied for several years with this subject and his first publication, in 1902, was devoted to thermal equilibrium, the statistical interpretation of temperature, entropy, and the equipartition of the energy principle. The second was concerned with irreversibility, and the third with fluctuations and new ways of determining Boltzmann's constant. He was apparently only partially aware of Maxwell's Kinetic Theory and Boltzmann's major publication of 1877, and there was a continuing confusion and disagreement about some of the intricacies of statistical methods.

Einstein's first paper to make any reference to Planck's radiation formula was published in 1904 when he wrote that Planck's formula had been derived using a statistics expression for the entropy of ideal gases. It was the beginning of his interest in quantum theory in the context of statistics, and his starting point was to use Maxwell–Boltzmann statistics to investigate **fluctuations** of energy about the mean in thermal equilibria. This led to an interest in the volume dependence of entropy and, as mentioned above, these topics were to influence his work for nearly a quarter of a century.

1905

This was the year in which Einstein, at the age of 26, advanced the course of physics in three ways that revolutionised scientific thinking.

Firstly, after Planck's work on black-body radiation, one of Einstein's famous papers of 1905 was the next most significant step in the development of quantum physics. In that paper, of March 1905, he developed and justified the concept of light actually being transmitted as discrete '**energy quanta**' (his term). Noting that optical phenomena such as diffraction, reflection, refraction, and dispersion are well accounted for by the wave model of light using the continuous functions of **Maxwell's Electromagnetic Theory**, he emphasised that such observations refer to time averages rather than instantaneous values. This was in contrast to observations of phenomena connected with the emission or transformation of light, as in the production of black-body radiation, the **photoelectric effect**, etc. These would be more readily understood if one assumed the energy of light to be discontinuously distributed in space. He noted that in the case of matter ('ponderable bodies'), energy is described by a summation over the constituent 'atoms and electrons'. With all this in mind, he reasoned that the energy of light spreading from, say, a point source is not continuously distributed but consists of his '**energy quanta**' that are 'localised at points in space' and which 'move without dividing and which can only be produced and absorbed as complete units'. [He later regretted referring to 'points' — in view of the **Uncertainty Principle** that came later (§9.4.2).]

Einstein used thermodynamic and statistical considerations to elaborate and justify this (§4.3) and he went on, in the same paper, to show that it led to a successful explanation of the **photoelectric effect** (see Appendix F), with a prediction, later confirmed experimentally, of a linear relationship between the 'stopping power' of photoelectrons and the frequency of the incident radiation. Four years later, he was going to show the need for a 'fusion of wave and quantum interpretations' of light (§4.4.1).

Secondly, in 1905, Einstein also published, in two papers, his theory of relativity (Einstein, 1905(d)(e)). This was later known as his **Special Theory of Relativity** (Chap. 2) to distinguish it from his **General Theory of Relativity** published in 1915 and dealing with gravity.

These two 1905 papers received little notice until nearly a year later, when Planck drew attention to their profound importance. Planck's main interest in science was to search for the 'absolute', and it is summed up in his scientific autobiography where he wrote that "... the Theory of Relativity confers an absolute meaning on a magnitude which in classical theory has only a relative significance: the velocity of light. The velocity of light is to the Theory of Relativity as the elementary quantum of action is to the quantum theory". He was the first to introduce the Special Theory of Relativity into quantum theory, showing in 1906 that h is a relativistic invariant (i.e. independent of reference frame).

The third of the great advances Einstein made in 1905 was his work on the application of fluctuations to an understanding of **Brownian Motion** (Einstein, 1905(b)(c) also 1906(a)). He accurately predicted how far molecular collisions in a liquid would push suspended pollen grains. It was a major step at a time when little was known about atomic and molecular structure. It also confirmed for him the correctness of **Boltzmann Statistics**.

In 1921, Einstein was given the Nobel Prize for physics for his "attainments in mathematical physics and especially for his discovery of the law for the **photoelectric effect**". In 1982, his biographer, Abraham Pais, rightly commented that "no one before or since has widened the horizons of physics in so short a time as Einstein did in 1905".

1906

This year saw an extremely important discussion of Einstein's 'light quanta' in relation to Planck's 'energy elements' (Einstein, 1906(b)).

After arriving at the concept of light quanta (and after producing his Special Theory of Relativity) Einstein returned to Planck's theory of black-body radiation, and in particular the latter's use of the equation (§3.2.2, Eqn. 3.9):

$$\rho(v, T) = \frac{8\pi v^2}{c^3} E(v, T), \qquad (4.1)$$

which had been the foundation of Planck's work. In this equation, $\rho(v, T)$ is the energy distribution function of radiation and $E(v, T)$ is the average equilibrium energy of resonators whose vibration frequency is v. Planck had derived this from purely Classical Mechanics and Electrodynamics. He had then found that he had to introduce what we can now call quantisation into the **resonator energy** $E(v, T)$ in order to obtain his famous radiation formula (Eqn. 3.34) that agreed with all the experimental data:

$$\rho(v, T) = \frac{8\pi v^2}{c^3} \frac{hv}{e^{hv/kT} - 1}. \qquad (4.2)$$

This step that Planck had taken was of course quite alien to Classical Theory. In contrast, Einstein, in his 1905 paper, arrived at the concept of the light quantum (with which he then very successfully explained the photoelectric effect) solely from thermodynamic and statistical considerations (§4.3) without it being invoked as it was in Planck's theory of black-body radiation.

To reconcile the two obviously successful quantisation concepts, Einstein proposed that Planck's initial classically-derived equation (Eqn. 4.1) must also be accepted as a valid quantum theory, though its foundation was a mystery when quantum effects are important. This was a major decision and detailed examination of Planck's derivation of the equation led him to conclude that "We must consider the (following) theorem to be the basis of Planck's radiation theory: the energy of a [Planck oscillator] can take on only those values that are integral multiples of hv; in emission and absorption the energy of a [Planck oscillator] changes by jumps which are multiples of hv" (Pais — see Bibliography). We had $E = nhv$, where $E = hv$ became known as **Einstein's Energy Equation**.

Einstein's acceptance of Planck's initial equation was a major step forward because it was no longer a hindrance to his thinking. In contrast, Planck's acceptance of Einstein's work, in which light was seen to be transmitted in quanta, was not so immediate because Planck still assumed that "what happens in the vacuum is rigorously described by Maxwell's equations".

1907

A quantum theory of **specific heats** was published which solved the anomaly in the classical treatment (§4.5). It was an early demonstration of the wider applicability of quantum theory that saw the birth of solid state quantum theory. In this paper, Einstein used his work on fluctuations (§4.4), at the same time showing that this also provided yet another derivation of Planck's radiation formula.

1909

Two profound papers by Einstein were published this year, between the presentation of which he moved from being a 'second-class technical expert' at the Bern Patent Office to become associate professor in Zürich. In these, he applied his 1904 work on energy fluctuations to black-body radiation and deduced that radiation consists of "moving point-like quanta with energy $h\nu$". This was the first mention of 'point-like' quanta. At the same time, he said that the theory suggested a "kind of fusion of the wave and the emission theory ... The wave structure and the quantum structure ... are not to be considered as mutually incompatible ..." (§4.4.1).

In these 1909 papers, Einstein also presented his first results concerning **momentum fluctuations** (§4.4.2). They fell short of explicitly recognising light quanta as having momentum: that did not happen until 1916 in his work on Brownian Motion. This is strange because, as is often commented, it could have already been deduced from his 1905 Special Theory of Relativity work. There, the energy of a particle is given (cf. Eqn. 2.13) by

$$E = \sqrt{p^2c^2 + \left(mc^2\right)^2},$$

which for a photon ($m = 0$) gives

$$E = pc.$$

Combined with Einstein's $E = h\nu$, this gives **photon momentum** as

$$p = \frac{h\nu}{c} = \frac{h}{\lambda}. \qquad (4.3)$$

1910–1925

A number of papers were published during these years by Einstein, dealing with various topics such as the thermodynamics of photochemical processes, further work on fluctuations (§4.4.2), and on specific heat theory (§4.5). The year 1915 saw the publication of his **General Theory of Relativity**.

In 1916–1917, Einstein returned again to black-body radiation, with three papers based on quite general hypotheses about the interaction between the radiation and the gas in the radiation enclosure. This work also gave a new derivation of Planck's radiation law. It involved no specific assumptions about the type of matter interacting with the radiation.

Einstein was here considering transitions between energy levels in an equilibrium system. He showed that, to be compatible with Planck's experimentally proven Radiation Law, a transition would also be associated, given the appropriate circumstances, with a quantum of radiation — in accord with Bohr's theory (which we describe in the next chapter). He had significantly established a bridge between black-body radiation and Bohr's theory of spectra. Because of the relevance of this work of Einstein to both topics, and its great subsequent relevance in Heisenberg's development of **matrix mechanics,** we deal with it separately, in Chap. 6, after looking at Bohr's theory of the atom.

Two of the papers in 1916–1917 had established more explicitly the association of momentum with light quanta than had his 1909 work. This was an extension of the earlier paper but it now dealt with how a Maxwell distribution of velocities is maintained in a molecular gas subject to radiation pressure. This linked it with **Brownian Motion** (on which he had already published an important work (Einstein, 1905(b)) and clearly showed the association of light with momentum i.e. $p = h\nu/c$.

However, a proper explanation of the electromagnetic field, especially black-body radiation, was still lacking, and the pioneering studies by Einstein (1909(a)(b) and 1917), Debye and others (such as Ehrenfest), were superseded in 1927 by Dirac and others with theories that became known as **Quantum Field Theory**. In that, particles (with any number of degrees of freedom) are represented by fields that have quantised normal modes of oscillation. The case where the photon is treated as the quantum 'particle' of the electromagnetic field — **quantum electrodynamics (QED)** — was developed, notably and independently, by Feynman, Tomonaga, and Schwinger to whom the 1965 Nobel Prize for physics was jointly awarded. The theory became highly successful in dealing with photons, electrons, and positrons, and their interactions, with 'profound consequences for elementary particle physics'.

Returning to the period we are dealing with here, during 1924–1925 Einstein published further work on the quantum theory of a molecular gas, and in 1925 his last application of the fluctuation theory led him to particle–wave duality for matter by an approach independent of that taken earlier by de Broglie. His '**gas theory**' was to be a powerful influence on Schrödinger in his approach to developing his wave mechanics (Chap. 7).

One of the most crucial experiments actually demonstrating more directly the reality of light-quanta and their having momentum — more directly than had the **photoelectric effect** — came in 1923 when Arthur Compton showed that the increase in wavelength of X-rays (i.e. their reduction in energy) that occurs when they are scattered by the loosely-bound outer electrons of light elements was in accord with the conservation of energy and mechanical momentum — as in the collision of two elastic spheres (Compton, 1923(a)). [Compton also considered the quantum aspects of diffraction; see Compton, 1923(b)] Compton was awarded the 1927 Nobel Prize in physics for the '**Compton effect**' (the award being shared that year with C. T. R. Wilson for his development of what we know as the '**Wilson Cloud Chamber**').

*   *   *

We have noted that Einstein's 1904 paper was the commencement of his use of statistical methods in investigating quantum matters, and which continued right through to 1925. His investigations of fluctuations in systems in thermal equilibrium drew his attention to the rôle of the volume-dependence of entropy. Although fluctuations featured in most of his subsequent work in this area, his first publication after the 1904 paper was an application of the volume dependence of entropy. This was the famous March 1905 paper that saw the birth of Einstein's '**light quanta**' and an explanation of the **photoelectric effect**. We will therefore look at the basis of that first (§4.3). Then we will turn to the basis of his uses of fluctuations in his subsequent papers (§4.4). He also used this statistical approach in several derivations of Planck's radiation formula and in his very important development of a theory of specific heats, that resolved the shortcomings of the classical **Dulong and Petit Rule**.

4.3. Volume Dependence of Entropy: Light Quanta and the Photoelectric Effect

It was Einstein's March 1905 paper that announced the 'revolutionary' hypothesis that light itself consists of '**energy quanta**' (or '**light quanta**')[b] and Einstein successfully used it to explain the **photoelectric effect** (see Appendix F). The hypothesis was based on obtaining, for radiation, an equation relating the change in entropy with the change in the volume occupied by the radiation, and comparing it with the

---

[b] Although Einstein introduced the terms '**energy quantum**' and '**light quantum**' in this 1905 paper, Planck had used the word '**quantum**' once in 1901 in referring to the charge on the electron as the '**elementary quantum of electricity**' — a term that was not adopted.

following well-known equation for gases:

$$S - S_0 = \frac{R}{N} \log \left(\frac{V}{V_0} \right)^n, \tag{4.4}$$

where N is Avogadro's number and R is the gas constant. (Not until 1909 would Einstein write k for R/N.)

This equation refers to a system such as an ideal gas or dilute solution, initially with volume V_0 and entropy S_0, containing what Einstein described as n movable points (e.g. molecules). If all the movable points are 'transferred' without any other change into a smaller volume V (part of V_0), the new state will have a different entropy, S, the difference being given by the above equation. (Einstein used his own way to arrive at the equation because of his continuing dissatisfaction with Boltzmann's method of counting '**complexions**'. However, Boltzmann's method was subsequently accepted and seen to be more widely applicable.)

To see how Einstein set up the analogous equation he wanted for radiation, we will proceed in two steps.

Step 1

It is assumed that, unless disproved by experiment, the observable properties of radiation are fully determined when the radiation density is given as some function of its frequency, $\rho(v)$ say. Radiation of different frequencies are assumed to be independent when there is no transfer of heat or work, and the entropy per unit volume can then be expressed as

$$S = \int_0^\infty \varphi(\rho, v) dv, \tag{4.5}$$

where φ is a function of ρ and v.

If the temperature of a unit volume is raised by dT, then the increase in entropy is given by

$$dS = \int_{v=0}^{v=\infty} \frac{\partial \varphi}{\partial \rho} d\rho \, dv. \tag{4.6}$$

Einstein reasoned in several ways that for black-body radiation, $\partial\varphi/\partial\rho$ is independent of ν and the above equation then becomes

$$dS = \frac{\partial\varphi}{\partial\rho}dE, \tag{4.7}$$

since dE is equal to the heat added i.e.

$$dE = \int_{\nu=0}^{\nu=\infty} d\rho\, d\nu.$$

As always in this context, we assume the process is reversible and again the equation

$$\frac{dS}{dE} = \frac{1}{T}, \tag{4.8}$$

is applicable (cf. Eqn. 3.11).

Therefore, comparing the above equations, we have

$$\frac{\partial\varphi}{\partial\rho} = \frac{1}{T}, \tag{4.9}$$

which we shall use later.

Step 2

To proceed, and reluctant to use Planck's radiation formula (Eqns. 3.34 and 4.2) because of its lack of a sound theoretical basis, Einstein turned to **Wien's Distribution Law** (Eqn. 3.8) namely,

$$\rho(\nu, T) = \alpha\nu^3 e^{-\beta\nu/T}, \tag{4.10}$$

which fitted experimental data well at large values of ν/T and which was based on the concept of gases radiating under thermal excitement (§3.1).

Equation 4.10 can be written as

$$-\frac{1}{\beta\nu}\log\left(\frac{\rho}{\alpha\nu^3}\right) = \frac{1}{T}$$

$$= \frac{\partial\varphi}{\partial\rho}, \quad \text{by Eqn. 4.9.} \tag{4.11}$$

Integrating

$$\int_0^\varphi d\varphi = -\frac{1}{\beta v} \int_0^\rho \log\left(\frac{\rho}{\alpha v^3}\right) d\rho,$$

$$\varphi = -\frac{1}{\beta v}\left[\rho \log\left(\frac{\rho}{\alpha v^3}\right) - \rho\right]_0^\rho$$

$$= -\frac{\rho}{\beta v}\left[\log\left(\frac{\rho}{\alpha v^3}\right) - 1\right]. \tag{4.12}$$

To move towards an expression for comparison with Eqn. 4.4 we consider the energy $E(v)$ of 'monochromatic' radiation with frequency between v and $v + dv$, i.e.

$$E(v) = \rho V_0 dv, \tag{4.13}$$

where V_0 is the volume occupied by the radiation.

The corresponding entropy, S_0, is then

$$S_0(v) = \varphi V_0 \, dv,$$

and substituting from Eqn. 4.12,

$$S_0(v) = -\frac{\rho}{\beta v}\left[\log\left(\frac{\rho}{\alpha v^3}\right) - 1\right] V_0 \, dv,$$

and using Eqn. 4.13,

$$S_0(v) = -\frac{E(v)}{\beta v}\left[\log\left(\frac{E(v)}{V_0 \alpha v^3}\right) - 1\right]. \tag{4.14}$$

As with the case of a gas at the beginning of this section, we consider compressing this radiation into a smaller volume V. The entropy, $S(v)$, is then

$$S(v) = -\frac{E(v)}{\beta v}\left[\log\left(\frac{E(v)}{V \alpha v^3}\right) - 1\right]. \tag{4.15}$$

The change in entropy this time is

$$S - S_0 = \frac{E}{\beta v} \log\left(\frac{V}{V_0}\right) \tag{4.16}$$

This can be put into the Boltzmann form of Eqn. 4.4, which we must remember normally refers to probabilities of 'particles', not radiation,

$$S - S_0 = \frac{R}{N} \log \left(\frac{V}{V_0} \right)^{NE/R\beta v} . \tag{4.17}$$

Comparison of this with Eqn. 4.4 gave Einstein his famous hypothesis — "Monochromatic radiation of low density (within the range of validity of Wien's radiation formula) behaves thermodynamically as though it consisted of a number of independent energy quanta of magnitude $R\beta v/N$".

It is interesting to remind ourselves that this had been arrived at using statistical mechanics and classical thermodynamics, *without any reference to the quantum idea that Planck had invoked in connection with his resonators*. Einstein described the approach he had used in his own paper (Einstein, 1905(a)) as '**heuristic**', i.e. a branch of logic using a method not necessarily correct or provable but aiding the discovery of truth — in other words, an inspired guess where chance favours the prepared mind.

Einstein had put

$$E = R\beta v/N, \tag{4.18a}$$

and at that stage he kept to this expression for the energy. However, with $\beta = h/k$ (from a comparison of Wien's law (Eqn. 3.8) with Planck's equation (Eqns. 3.35 and 4.2), and $k = R/N$ as later understood in Planck's $S = k \log W$ (Eqn. 3.20) we have the familiar

$$E = hv, \tag{4.18b}$$

which, as noted in the previous section, became known as **Einstein's Energy Equation**, although 'h' did not appear in his 1905 paper.

Having enunciated his hypothesis, Einstein went on to say "If the entropy of the monochromatic radiation depends on volume as though the radiation were a discontinuous medium consisting of energy quanta of magnitude $R\beta v/N$, the next obvious step is to investigate whether the laws of emission and transformation of light are also of such a nature that they can be interpreted or explained by considering light to consist of such energy quanta".

In the next section of that March 1905 paper, Einstein did indeed successfully use this new concept of 'light quanta' to provide for the first time an explanation of some details of the **photoelectric effect** (Appendix F). It had been found that the photoelectron energies were surprisingly independent of the intensity of the illumination and that whilst their energies were dependent on the frequency of the illumination, there was a critical frequency, dependent on the metal of the detector terminal, below which no electrons were emitted.

Einstein's light quanta gave a direct explanation of these findings. The absorption of a light quantum by an electron in the metal surface would be converted into energy that may be sufficient to eject it from the atom. Its maximum energy would be given by $R\,\beta v/N$ according to Einstein's theory, and therefore it would be the frequency of the light, not its intensity, that determines the energy of the photoelectrons: increasing the intensity would simply result in more photoelectrons of the same energy. However, the kinetic energy of the photoelectrons would be less than the energy absorbed because work P, say, is required to remove them from the metal. The equation for their energy would therefore be of the form

$$KE_{\max} = (R\beta v/N) - P.$$

Hence, also as observed, there is an energy below which an electron would not be emitted.

This was strong evidence that the energy of radiation is actually conveyed (contrary to Planck's belief) by small individual units, 'light quanta'. As Einstein's biographer Pais put it, the genius of the light-quantum hypothesis had lain in the intuition for choosing the right piece of experimental input and the right, utterly simple, theoretical ingredients.

Einstein went on in his paper to point out that this theory of the photoelectric effect had further consequences. These were confirmed in due course in extensive experimental studies by R. A. Millikan (1916) who was awarded the 1923 Nobel Prize for physics for his work in 1911 on establishing the **electron** as a fundamental unit of charge, and for his work on the photoelectric effect. He was also renowned for his work on **cosmic rays** (which he so-named).

4.4. Fluctuations

4.4.1. *Energy fluctuations*

For his investigation in 1904 of energy fluctuations, Einstein considered a small volume in thermal equilibrium, at temperature T, with a large volume that acts as a thermal reservoir to accommodate fluctuations about the mean in the small volume. We assume that the overall average energy of the radiation is the sum of the average energies of each independent degree of freedom. For the sth degree, the average energy is given by the **Maxwell–Boltzmann function** (cf. Appendix B):

$$\langle E_s \rangle = \frac{\int E_s e^{-E_s/kT} dp_s \, dq_s}{\int e^{-E_s/kT} dp_s \, dq_s}, \tag{4.19}$$

where the integrals may, if appropriate, be replaced by summations. The total overall average energy is expressed as

$$\langle E \rangle = \sum_s \langle E_s \rangle. \tag{4.20}$$

A fluctuation about a mean is specified by its 'mean square', which in the present example is given by the average of the square of the difference between the actual total energy $\sum_s E_s$ and the total average energy $\sum_s \langle E_s \rangle$. Denoting the mean square by $\langle \Delta E^2 \rangle$, we have

$$\langle \Delta E^2 \rangle = \left\langle \left(\sum_s E_s - \sum_s \langle E_s \rangle \right)^2 \right\rangle$$

$$= \sum_s \sum_{s'} \langle (E_s - \langle E_s \rangle)(E_{s'} - \langle E_{s'} \rangle) \rangle. \tag{4.21}$$

It is assumed that we are dealing with mutually-independent degrees of freedom, in which case the cross-terms in the above products are zero and the equation becomes

$$\langle \Delta E^2 \rangle = \sum_s \left(\langle E_s^2 \rangle - \langle E_s \rangle^2 \right). \tag{4.22}$$

With $\langle E \rangle$ as a function of T and using Eqn. 4.19, Einstein obtained

$$\langle \Delta E^2 \rangle = kT^2 \frac{\partial \langle E \rangle}{\partial T} P (= kT^2 C_V). \qquad (4.23)$$

This equation proved to be a very important 'theorem' [with another derivation in 1909 (see Klein, 1964)]. It holds regardless of whether or not the radiation obeys Maxwell's electromagnetic equations, and whether or not the energy has a quantum nature. The only assumptions made were Boltzmann's Equipartition Principle, the mutual independence of the degrees of freedom of radiation (cf. colours travel separately), and the requirement that the spectral distribution of energy is only a function of temperature (cf. with a gas, the density and degrees of freedom would also fluctuate with temperature about the average, and the treatment would not be valid).

Einstein regarded $\langle \Delta E^2 \rangle$ as a measure of the thermal stability of a system — the larger its value the lower the stability — and he introduced a criterion for fluctuations to be large,

$$\xi = \frac{\langle \Delta E^2 \rangle}{\langle E \rangle^2} \approx 1. \qquad (4.24)$$

Using the **Stefan–Boltzmann law** (Eqn. 3.4) for $\langle E \rangle$ gives

$$\langle E \rangle = \sigma V T^4, \qquad (4.25)$$

where V is the volume (Eqn. 3.4 was for unit volume). With Eqn. 4.23, we therefore have

$$\langle \Delta E^2 \rangle = kT^2 4 \sigma V T^3, \qquad (4.26)$$

and therefore

$$\xi = \frac{4k}{\sigma V T^3}. \qquad (4.27)$$

With Wien's 'Displacement Law' expressed as in Eqn. 3.6, the above equation becomes

$$\xi \propto \frac{\lambda_{max}^3}{V}. \qquad (4.28)$$

It was this that drew Einstein's attention to the volume dependence of thermodynamic functions, which as we have seen led to his historic paper of 1905.

This fluctuation work also led Einstein in 1909 to find that the statistical approach gave the first intimation of **wave–particle duality** of radiation. In this he applied Eqn. 4.23 to the case where it is supposed that, in the original set-up, only thermal radiation with a frequency range between v and $v + dv$ in the small volume V can have an exchange with the surrounding reservoir volume. In Eqn. 4.22, $\langle E \rangle$ would then refer just to this selected range and can be written as

$$\langle E \rangle = \rho V dv, \tag{4.29}$$

where ρ is the energy density for frequency v. Equation 4.23 then becomes

$$\langle \Delta E^2 \rangle = kT^2 V dv \frac{d\rho}{dT}, \tag{4.30}$$

and with ρ from Planck's law (Eqn. 4.2) i.e.,

$$\rho(v, T) = \frac{8\pi h v^2}{c^3} \frac{1}{e^{hv/kT} - 1}.$$

This gives mean square energy as

$$\langle \Delta E^2 \rangle = \left(hv\rho + \frac{c^3}{8\pi v^2} \rho^2 \right) V \, dv. \tag{4.31}$$

This suggested that there are two different, additive causes producing fluctuations. For low frequencies, and high temperatures, only the second term in Eqn. 4.31 would be significant, and this would correspond to the classical relationship e.g. Planck's law in the limit and the Rayleigh–Jeans law (Appendix D). At the other extreme, of high frequencies and low temperatures, the first term dominates and Einstein commented that this would "result in fluctuations if radiation were to consist of independently moving particle-like quanta with energy hv".

This was the basis for Einstein's suggestion in 1909 (referred to in §4.2) about the wave–particle duality of radiation.

4.4.2. *Momentum fluctuations*

Einstein's second paper on this topic in 1909 concerned momentum fluctuations, though falling short of explicitly recognising light quanta as having momentum (as mentioned in §4.2, that came in 1916 in connection with his work on Brownian motion).

He considered the effect of radiation pressure on a hypothetical, perfectly reflecting, mirror that is free to move in the direction perpendicular to its own plane. The mirror is in the usual black-body radiation vessel containing ideal gas at low pressure at a temperature fixed by the temperature of the walls of the vessel. Gas molecules collide with the mirror and set it into irregular motion rather like Brownian motion. Because of this motion, the radiation pressure on either side of the mirror will be unequal. The unbalanced radiation pressure opposes the motion of the mirror ('radiation friction') and this increases as the velocity of the mirror increases. The kinetic energy of the mirror becomes converted into energy of the radiation field, and this would bring the mirror to rest were it not for the fluctuations of the radiation pressure. Using Planck's formula for the radiation energy density distribution, a long calculation (part of which Einstein published with L. Hopf a year later in 1910) gave an expression for the mean square momentum fluctuations, and it was similar to Eqn. 4.31. As before, the two terms could be identified as arising individually from the Wien and Rayleigh–Jeans limits of Planck's radiation formula, the first interpretable if the radiation were particle-like and the second if associated with a wave phenomenon.

These papers of 1909 clearly suggested the need for a new approach to the theory of radiation. They contributed to Einstein's growing conviction of the necessity to consider momentum exchanges as well as energy changes in any theory of the interaction between radiation and matter.

As mentioned in §4.2, in 1916–1917 Einstein made further progress in the light of Bohr's new theory of the atom, and this is dealt with separately in Chap. 6 because of its relevance to both Planck's radiation law and Bohr's theory.

4.5. Statistics: Planck's Radiation Formula and Specific Heat Theory

In the previous sections, we have outlined how Einstein's 1904 study of fluctuations led to his work on the volume dependence of entropy and how in 1909 that led to the suggestion of **wave–particle duality**.

The statistical basis of that early work also led to his independent derivation of Planck's radiation formula (Eqn. 4.2) and to his theory of specific heat (1907). For this he expressed the average energy of a system as a function of frequency v and temperature T in the following fashion:

$$\langle E(v, T) \rangle = \frac{\int Ee^{-E/kT}w(E, v)dE}{\int e^{-E/kT}w(E, v)dE}. \tag{4.32}$$

Here, in the numerator, the energy E for a given v and T is multiplied by its probability [the exponential as in Appendix B, (Eqn. B.3)] and by its weighting $w(E, v)$ in the energy distribution (cf. **'density of states'**), and then integrated over all values of E. The denominator is simply for normalisation.

The corresponding expression Planck had used for this average equilibrium energy of a harmonic oscillator (Eqn. 3.33) in his final radiation formula (Eqn. 3.34) is

$$\langle E(v, T) \rangle = \frac{hv}{e^{hv/kT} - 1}. \tag{4.33}$$

Einstein introduced quantisation into Eqn. 4.32 by proposing that the energy states be restricted to having values of E given by

$$E = nhv, \tag{4.34}$$

where $n = 0, 1, 2 \ldots$ (§4.2, see the subsection on the year 1906).

Strictly, E would have values given by

$$nhv < E < nhv + \alpha, \quad \alpha \ll nhv,$$

with

$$\int_{nh\nu}^{nh\nu+\alpha} w\nu dE = \text{constant},\qquad(4.35)$$

for all n.

With this specification of E, the numerator in Eqn. 4.32 takes the form (omitting constants) of a summation,

$$h\nu e^{-h\nu/kT} + 2h\nu e^{-2h\nu/kT} + 3h\nu e^{-3h\nu/kT} + \cdots,$$

and the denominator is

$$1 + e^{-h\nu/kT} + e^{-2h\nu/kT} + \cdots$$

Using the binomial theorem for both readily gives Planck's expression (Eqn. 4.33) without recourse to any of Planck's assumptions. (Jeans later gave a similar derivation.)

In his 1907 paper, Einstein then turned to the problem with **specific heat theory** (§3.1) where the classical value kT for the mean energy of an atom failed to give results in accordance with experiments at low temperatures. He used a very simple model of a solid represented by a three-dimensional lattice with atoms oscillating independently, isotropically, and harmonically, with a single frequency ν about their equilibrium position. Instead of the mean energy of a gram atom being the classical $3NkT$, he expressed it, using Eqn. 4.33, as

$$3NkT\frac{h\nu/kT}{e^{h\nu/kT} - 1},\qquad(4.36)$$

where N is Avogadro's number. (This still gives the classical value for high temperature.)

The specific heat $C_V(=dE/dT)$ is then

$$3Nk\frac{x^2 e^x}{(e^x - 1)^2},\qquad(4.37)$$

where $x = h\nu/kT$.

For a given solid, the value of the frequency in the above expression was obtained empirically from a value of C_V that was known at a suitable temperature to enable values of C_V to then be calculated at low temperatures for comparison with experiment.

The results were in remarkably good agreement with experimental data (of the form shown in Fig. (3.1)) in view of the simplifications and assumptions involved.

Einstein developed the specific heat theory further but it was treated more fully by Debye, and by Born and Kármán, with the benefit of the information about the structure of the solid state that became available in the years after the Laue experiment of 1912. Notes on Debye's treatment are given in Appendix E because his approach influenced Schrödinger in developing wave mechanics.

*   *   *

Thus, we have seen that the quantum was firmly established in radiation theory by 1913 when Bohr produced his theory of the role of the quantum in the atom, to which we turn in the next chapter. In 1916, Einstein returned to the statistical aspects of black-body radiation and not only provided another way to Planck's radiation formula but also dealt with a problem concerning the intensities of atomic spectra in the Bohr theory. We refer to it again in the next chapter but since it is common to both, we give an outline separately in Chap. 6.

Chapter 5

The Quantum in the Atom: Optical Spectra

5.1. Introduction

The so-called '**Old Quantum Theory**' introduced by Planck and Einstein had arisen from a failure of classical physics to account for **black-body radiation**. The rôle of quantisation was about to solve, more far-reachingly than could have been expected at that time, the problem of another breakdown in classical theory.

Rutherford's 1911 planetary model of the structure of the atom, clearly evident experimentally, was based on Newton's laws of motion and Coulomb's law of electric force. However, an electron in an orbit round a positive nucleus is an accelerating charge and, according to **Classical Electrodynamics**, it would radiate energy and therefore spiral into the nucleus, which was obviously not the case.

Niels Bohr solved the problem in 1913. He reasoned that, like the quantum nature of electromagnetic radiation, perhaps there is quantisation of mechanical energy too. After all, if an excited atom emits light in quanta of energy $h\nu$, as had been established by Planck and Einstein, the mechanical energy of the atom must at the same time be reduced by the same amount. Atomic spectra provided the explanation. Atomic spectra are observed experimentally, with e.g. a prism, as discrete 'lines' which indicated that there are different states of an atom that are somehow associated with definite energies. Bohr found an explanation of this that was supported experimentally. It was going to involve the concept of '**stationary states**', in which electrons occupy orbits without loss of energy, their energy only changing when they jump from one orbit to another in the process of emitting (or absorbing) radiation.

Various simple optical emission line spectra had been studied experimentally by the end of the 19th century and an *empirical* formula for the hydrogen spectrum had been found by Balmer. Rydberg put this into a general form as follows, written here in terms of frequency v:

$$v = R\left(\frac{1}{m^2} - \frac{1}{n^2}\right), \tag{5.1}$$

where m and n are integers and R is the **Rydberg Constant**. For the hydrogen spectra that Balmer had observed, $n = 2$ and $m = 3, 4, 5, \ldots$

Bohr found that by introducing Planck's quantum concept into the mechanics of orbital electrons, he could explain the above empirical formula with m and n than having a quantum interpretation.

Equally important, Bohr also showed that there is a region of scale (in the present case, large m and n) where the new quantum laws merge into the classical laws. This came to be known as his '**Correspondence Principle**' and it was one of his most important contributions because it formed the basis of Heisenberg's **Matrix Mechanics** (Chap. 9).

Bohr's detailed and sometimes convoluted deliberations on these matters led to a number of publications, especially during 1913–1915. A number of '**postulates**' emerged, sometimes retrospectively, which are variously quoted by different authors, and Bohr gave several derivations to justify his quantisation ideas. For example, one can read that he postulated that the angular momentum of an orbital electron is quantised. However, this is not the only manifestation of the quantisation and we shall pursue the route in which he quantised the orbital energies because this leads more clearly to his establishment of the **Correspondence Principle**. Furthermore, this approach relates atomic energy quanta to the orbital electron rotation frequency in a way analogous to how Planck had related resonator energy quanta to resonator oscillation frequency.

We need first to recall some details of the Classical Mechanics of orbital electrons to appreciate how the quantum ideas were incorporated.

5.2. Classical Mechanics of an Electron in a Circular Atomic Orbit

The orbital energy, A, of an electron with charge e, mass m and velocity v, in a circular orbit of radius r about a nucleus with atomic number Z, is the sum of its kinetic and potential energies:

$$A = \frac{1}{2}mv^2 - \frac{Ze \cdot e}{r}. \tag{5.2}$$

(The negative sign follows the convention that the electron has zero potential energy when it is at infinity.)

The radial electrostatic force between the electron and the nucleus is $Ze \cdot e/r^2$ and this is balanced by the centripetal force (Newton's second law), $m\omega^2 r$,

$$\frac{Ze \cdot e}{r^2} = m\omega^2 r = \frac{mv^2}{r}, \tag{5.3}$$

where ω is the angular frequency.

Equation (5.2) therefore becomes

$$A = \frac{Ze^2}{2r} - \frac{Ze^2}{r}$$
$$= -\frac{Ze^2}{2r}$$
$$= -\frac{mv^2}{2}, \tag{5.4}$$

i.e. the total energy is the negative of the kinetic energy (or half the potential energy).

Bohr took the numerical value here to represent the 'binding energy' (or 'ionisation energy') of the electron in its orbit, i.e.

$$W = \frac{Ze^2}{2r} = \frac{mv^2}{2}. \tag{5.5}$$

The angular frequency of the orbital electron is given in terms of W from Eqns. 5.3 and 5.5,

$$\omega^2 = \frac{8W^3}{mZ^2e^4}. \tag{5.6}$$

Converting to rotation frequency $f = \omega/2\pi$ for comparison with spectral frequencies later,

$$f^2 = \frac{2W^3}{\pi^2 mZ^2 e^4}. \tag{5.7}$$

Strictly, m is the 'reduced mass' of the electron due to the effect of the nucleus itself having a small associated orbit of rotation. Also, relativistic effects have to be taken into account. Details can be omitted because they are irrelevant here and would detract unnecessarily from the main storyline.

Before moving on to the introduction of the quantum, it is interesting to note that if we square the second expression for A in Eqn. 5.4 and divide it by the third expression we find the orbital energy is given by

$$A = -\frac{1}{2}\frac{Z^2 me^4}{(m\upsilon r)^2}, \tag{5.8}$$

which foreshadows finding that quantising the orbital energy is concomitant with quantising the angular momentum (we return to this in §5.3).

5.3. The Quantum is Introduced: The Correspondence Principle

Three important 'postulates' (variously expressed and given with others by Bohr) underlie the quantisation. The essentials for us here are shown below.

(i) An electron can only occupy certain orbital states and they do so without radiating — they are 'stationary states'.

(ii) The dynamic equilibrium of an electron whilst in a stationary state is governed by the laws of classical mechanics.

(iii) An electron can jump, in a way not allowed by classical mechanics, from one stationary state to another, accompanied by the emission (or absorption according to the direction of energy change) of radiation of frequency, v, determined by

$$A_2 - A_1 = hv, \tag{5.9}$$

where h is Planck's constant. This is Bohr's 'frequency condition'.

[It is important to note that there is no assumption that v in Eqn. 5.9 is the frequency, f, of the orbital rotation as given by Eqn. 5.7.]

* * *

Bearing in mind how Planck had related resonator energy to its frequency, Bohr assumed that the binding energy, W, of an electron in an orbit was related to its orbital rotation frequency, f, by putting $W \propto hf$. Furthermore, he suggested that two different stationary states would have binding energies differing by multiples of hf,

$$W_n = \alpha nhf \tag{5.10}$$

where $n = 1, 2, 3, \ldots$ is a 'quantum number' and α is a constant. The latter was determined by invoking the Correspondence Principle as follows.

Substituting W_n from Eqn. 5.10 into Eqn. 5.7 gives

$$f_n = \frac{\pi^2 m Z^2 e^4}{2\alpha^3 n^3 h^3}, \tag{5.11}$$

whence, by Eqn. 5.10, we have

$$W_n = \frac{\pi^2 m Z^2 e^4}{2\alpha^2 n^2 h^2}. \tag{5.12}$$

Using this for the A's in Eqn. 5.9 for the nth and $(n + 1)$th orbitals gives

$$v = \frac{\pi^2 m Z^2 e^4}{2\alpha^2 h^3} \left(\frac{1}{n^2} - \frac{1}{(n + 1)^2} \right). \tag{5.13}$$

For very large n, this approaches

$$v = \frac{\pi^2 m Z^2 e^4}{2\alpha^2 h^3} \left(\frac{2}{n^3} \right), \tag{5.14}$$

and as $n \to \infty$ the energy difference between successive states tends towards being continuous. This is, therefore, entering the classical realm where the frequency v of the radiation is the orbital's rotation frequency f. Equating Eqns. 5.11 and 5.14 gives $\alpha = \frac{1}{2}$ and we have

$$\left. \begin{array}{l} f_n = \dfrac{4\pi^2 m Z^2 e^4}{n^3 h^3}, \\[3mm] A_n = -\dfrac{2\pi^2 m Z^2 e^4}{n^2 h^2} \\[3mm] \text{and} \quad W = \dfrac{nhf}{2}. \end{array} \right\} \tag{5.15}$$

Applying this expression for A_n to Eqn. 5.9 for hydrogen ($Z = 1$) gives

$$A_2 - A_1 = h v_{2,1} = \frac{2\pi^2 m e^4}{h^2} \left(\frac{1}{n_2^2} - \frac{1}{n_1^2} \right), \tag{5.16}$$

which gave good agreement with the experimentally-obtained value for the Rydberg constant [R in Eqn. 5.1]. This vindicated Bohr's hypotheses and it was regarded as the greatest agreement between theory and experiment since Maxwell derived the speed of light from purely theoretical calculations in his **Electromagnetic Theory**.

[Most texts give another of Bohr's approaches to quantisation, which is simpler but does not so clearly reveal the 'correspondence' to Classical Mechanics where the latter applies. In that approach, it is **angular momentum** that is **postulated to be quantised**, and Bohr expressed it as

$$m v r = \frac{nh}{2\pi}. \tag{5.17}$$

With Eqns. 5.3 and 5.4, this readily gives

$$A_n = -\frac{2\pi^2 Z^2 m e^4}{n^2 h^2}, \tag{5.18}$$

which is the same as what we obtained using the other starting point (Eqn. 5.15).]

A point to note that we shall use later (Chap. 8) concerns frequencies due to combinations of transitions. We have seen that a transition from a stationary state energy level A_n to level A_k results in the emission of a quantum of energy $A_n - A_k$ and whose frequency is

$$v(n, k) = (A_n - A_k)/h \qquad (5.19)$$

(Eqn. 5.9). Similarly for a transition from $k \to m$, we have

$$v(k, m) = (A_k - A_m)/h. \qquad (5.20)$$

For a transition from A_n to A_m, we would have an emission of frequency $v(n, m)$ given by

$$\begin{aligned} v(n, m) &= (A_n - A_m)/h \\ &= v(n, k) + v(k, m). \end{aligned} \qquad (5.21)$$

which means that if frequencies $v(n, k)$ and $v(k, m)$ occur in a spectrum, then the sum of these can also occur.

Such combinations became known as the 'Rydberg–Ritz Combination Rule', first recognised by Rydberg and first used by Ritz.

Ensuring that the quantum laws 'correspond' to the Classical laws in the region where the latter hold (as used above) was developed and elaborated over time by Bohr, although as the 'Correspondence Principle' it is usually given the date 1923. The essence of the idea was in fact used originally by Planck in developing the correct form of his radiation formula (§3.2.3). We return to it in Chap. 8 in its rôle in contributing to the basis of Heisenberg's **matrix mechanics**.

By 1916, Sommerfeld had extended Bohr's atomic model to deal with elliptical orbits and with elements containing more than one electron (see Appendix C, §C.3). He also took relativistic effects into account. There were now three **quantum numbers** — corresponding to the diameters, eccentricities, and spatial orientations of the electron orbits (later, in wave mechanics, they specified the nodes in the three-dimensional wave functions).

In dealing with the increase in atomic volume through the Periodic Table, Wolfgang Pauli in 1924 found it necessary to propose that only two electrons can occupy any given quantum state described by the three quantum numbers. This became known as the **Pauli Principle** (or **Pauli Exclusion Principle**) for which Pauli was awarded the 1945 Nobel

Prize for physics. With further developments by others, this provided a proper foundation for the Periodic Table, hitherto only constructed empirically by Mendeleev. Chemistry was put on a firm footing.

With this greater knowledge of atomic structure, spectroscopy became a huge subject in its own right, and it was one of the developments there that we need to note because of its importance in the subsequent development of quantum theory.

Anomalies were occurring in the interpretation of the additional line-splitting in the **Zeeman Effect** — the splitting of spectral lines by strong magnetic fields — discovered in 1896. (Zeeman shared the 1902 Nobel Prize for physics with H. A. Lorentz for their research on the effect of magnetism on radiation phenomena.) Various reasons were suggested, notably by Goudsmit and Uhlenbeck in 1925, who proposed that the effect was due, not to some further aspect of electron orbits, but to the electron itself. Perhaps the small, electrically charged electron spins like a top. It would have angular momentum and a magnetic moment, and the orientation of its 'spin', relative to the plane of its orbit, might account for the effect. The suggestion worked out well except that the angular momentum of the spinning electron had half-integer quantum numbers.

There had already been possible explanations of line-splitting effects, and the above tied in well with various suggestions (including **half-integer quantum numbers** and a fourth quantum number) already variously mooted by Compton, Kronig, Heisenberg, Landé, and especially Pauli. 'Spin' was to become recognised as a **fourth quantum number**.

The discovery of spin led to the **Pauli principle** being modified to state that only two electrons with 'opposite' spins can occupy any given quantum orbit.

Unfortunately, the term '**spin**' continued to be used even though it did not refer to spin in the literal sense (proved experimentally years later by Bohr). It has no classical meaning at all and occurs only in quantum interactions. Much of its mystery was resolved by Dirac in 1928.

Pauli contributed far more to the development of quantum physics than is apparent. At different times, he collaborated with the principal workers in the field, contributed much in discussions, and his sharply critical views were greatly valued and respected. His name otherwise appears mainly in connection with work leading to a quantum mechanical theory of electrons in metals, and also in nuclear physics where in 1931 he predicted the existence of a particle to explain the observation that some of the beta rays emanating from radioactive nuclei have less than the expected energy. His proposal, that the energy was partly carried away by a new subatomic particle, was developed further by Fermi who in 1932 named the particle the 'neutrino', and it was observed experimentally in 1956 by **Reines and Cowan**. Like Pauli, Fermi's contributions in physics were numerous. He was awarded the 1938 Nobel Prize for physics for the discovery of new radioactive elements produced by neutron irradiation, and for the discovery of reactions induced by slow neutrons.

One feature of Bohr's theory is of considerable relevance in present-day **cosmology**. If we turn back to the classical expression (Eqn. 5.8) for the energy of an electron (mass m, velocity v) in a circular orbit in the hydrogen atom ($Z = 1$) viz.:

$$A = -\frac{1}{2}\frac{me^4}{(mvr)^2},$$

and insert the quantisation of the angular momentum (Eqn. 5.17):

$$mvr = \frac{nh}{2\pi},$$

we can express A as

$$A = -\frac{1}{2}mc^2\left(\frac{2\pi e^2}{hc}\right)^2\frac{1}{n^2}. \tag{5.22}$$

The expression in brackets is of great interest. For use with SI units, it is $e^2/2\pi\varepsilon_0 ch$, where ε_0 is the permittivity of free space (originally defined as unity in the CGS system). Dimensionless, and universally given the symbol α, it has a value close to 1/137. Regarded as one of the fundamental constants of physics, it has long been reckoned to imply a deeper connection between the three entities e, c and h. Various ways of interpreting it have been pointed out — one is that it is a measure of the strength with which an electron binds in an atom. In that way, it appears in the more detailed expressions developed by Sommerfeld for spectra where it controls the fine-structure line-splitting. For that reason Sommerfeld named it the '**fine-structure constant**'.

Starting in 1999, there have been reports of evidence that α has not remained constant over the life of the Universe. This has emerged from measurements of the absorption spectra of distant **quasars** compared with present-day values. Whilst more accurate confirmation is awaited, the possibility of one or more of what have hitherto been regarded as universal constants changing with time is of profound importance.

*   *   *

Bohr's theory did not provide a way for deducing the *relative intensities* of the lines in a spectrum. However, in 1916 Einstein published a paper dealing with **transitions** between states in the context of **blackbody radiation,** and as noted in §4.2, he showed that to be compatible with Planck's law, a transition would also be associated with a quantum of radiation in accord with Bohr's '**frequency condition**', Eqn. 5.9. This had established a remarkable bridge between blackbody radiation and Bohr's theory of spectra and we therefore deal with it separately in Chap. 6. Bohr tried to link the two using his **Correspondence Principle** and to resolve the problem of intensities. Though not entirely successful, he did show that there is a close relation between the frequencies of the Fourier components of motion of a **non-linear oscillator** and the corresponding frequencies emitted according to his quantum theory. This model became very relevant as a springboard for Heisenberg's setting up of a completely new mechanics (Chap. 8).

Bohr's atom with its stationary states and its '**frequency condition**' was an unexplained contradiction of classical theory, but despite that, Bohr believed that his **Correspondence Principle** was the main link between classical mechanics and quantum mechanics. Also, he was still resisting the '**light quantum**' concept introduced by Einstein in 1905 (§4.2).

In his resistance to the concept of the light quantum (photon), Bohr published a paper in 1924 with Kramers and J. Slater (the latter providing its key starting point). The '**BKS Suggestion**' (a 'suggestion' because it contained no new research as such) attempted to dispose of the need for the concept, so preserving the 'continuous' picture of light (just as Planck had retained a belief in black-body radiation being transmitted as a wave motion, despite his concept of '**energy elements**' in its generation).

Disproved by the Compton Effect work (§4.2), the BKS Suggestion was a last stand of the '**Old Quantum Theory**'. It had played its part though, because it was the first attempt to tackle the **wave–particle duality** problem, and its stimulation of later work was acknowledged as such by Heisenberg in 1955.

The disagreement between Bohr and Einstein about the photon was to last a long time, but photon–wave duality was later to become central to Bohr's **Principle of Complementarity** (Chap. 9). Such is the way science progresses, with every step playing its part.

Bohr was awarded the 1922 Nobel Prize for physics for "his study of the structure of atoms and of the radiation which emanates from them".

Chapter 6

Einstein's Transition Probabilities: Bohr's Theory and Planck's Law

In 1911 Einstein despaired of making further progress with the question of the reality of quanta and turned his attention to relativity, leading in 1915 to his **General Theory of Relativity.** Then he returned to his interest in black-body radiation saying later that 'a splendid light' had dawned on him about the absorption and emission of radiation. During 1916–1917 he produced three very important papers which, in general terms, dealt with the statistics of transitions between stationary states in relation to the interaction of radiation with matter. It made no special assumptions about the nature of the 'objects' interacting with the radiation: Einstein called them 'molecules' but with no special meaning in mind (and not required to be identified with Planck's oscillators). This led to a completely independent derivation of Planck's radiation formula and it also established a link between black-body radiation and **Bohr's theory of spectra.** His work in these papers also convinced him at last that a light quantum has **momentum.** (We showed in §4.2 how this was evident in his earlier work.) He apparently regarded this result as more important than his derivation of Planck's radiation law.

Because these developments became relevant as a springboard for Heisenberg and Born in setting up the completely new '**matrix mechanics**' (Chap. 8) that played a very important rôle in the development of quantum physics, we will briefly outline the relevant parts of this work.

Bohr's atom had by then been established, including the 'frequency condition' (Eqn. 5.9), for the frequencies emitted (or absorbed) in transitions between energy states. Einstein reasoned that the black-body radiation spectrum could therefore now be interpreted in such terms and it should lead to Planck's formula. He showed that it did.

Einstein considered jumps between two states, an upper (excited) state 2 and a lower state 1. The probability of N_2 'atoms' in state 2 dropping down (i.e. emission of radiation) to state 1 in time dt would be $AN_2\,dt$, where A is a constant. Similarly, N_1 atoms in state 1 can absorb radiation to rise to the excited state 2, but unlike emission, the probability of this would depend on the density of radiation, u_v say, having the frequency required for the transition. He assumed that the probability of N_2 atoms changing to level 2 in dt by absorption would be Bu_vN_1dt where B is a constant.

Large A would correspond in Bohr's theory to a strong emitted spectral line, and large B to a strong absorption line.

As u_v would increase with temperature, one would then have

$$\text{absorption} \gg \text{emission},$$

i.e. $$Bu_vN_1dt \gg AN_2dt,$$

because A and B are constants and at high temperatures $N_1 \approx N_2$.

Einstein therefore assumed there must be an additional probability of spontaneous emission ('**induced emission**') and he suggested that it was proportional to the radiation density. It would of course also have to be proportional to the probability B of the excited state. Hence, for equilibrium he wrote

$$Bu_vN_1 = (A + Bu_v)\,N_2,$$

i.e. $$\frac{N_2}{N_1} = \frac{Bu_v}{A + Bu_v}. \tag{6.1}$$

Using the Maxwell–Boltzmann distribution for probabilities,

$$\frac{N_2}{N_1} = e^{-(E_2 - E_1)/kT}$$

$$= e^{-hv/kT}, \tag{6.2}$$

where E_1 and E_2 are the energies for the two states, and for the transition at frequency v, the Bohr frequency condition (Eqn. 5.9) gives $E_2 - E_1 = hv$.

Equating Eqns. 6.1 and 6.2,

$$Bu_v = (A + Bu_v) e^{-hv/kT}$$

i.e.

$$u_v = \frac{A}{B} \frac{1}{e^{hv/kT} - 1}, \tag{6.3}$$

which is Planck's law (Eqn. 3.35) if

$$\frac{A}{B} = \frac{8\pi hv^3}{c^3}. \tag{6.4}$$

Einstein had therefore derived a relation between A and B and, importantly, by this method of derivation had shown that the ideas at that time about transition probabilities between stationary states were compatible with classical thermodynamics. It was clear that the A's and B's had fundamental relevance, for example in determining intensities of line spectra, which had not been possible in Bohr's theory.

In this way Einstein had formulated statistical rules regarding the occurrence of radiation transitions between stationary states, assuming only that absorption and emission occur with a probability proportional to the intensity of the radiation. It had also suggested, as we have seen above, that spontaneous transitions occur at some *a priori* rate, regardless of external influence. Concerning the latter, Einstein drew attention to the analogy with the well-known laws that govern transformations in radioactive substances (Rutherford and Soddy, 1902).

Bohr used this in an attempt, combined with his **Correspondence Principle**, to resolve the problem of calculating the intensities of spectra — not possible with his orbital theory. Though not entirely successful, he found a close connection between the frequencies of the Fourier components of a near-linear oscillator and the corresponding frequencies emitted/absorbed according to his quantum theory of orbital stationary states.

We saw in §5.3 that Bohr had used the fact that for large quantum numbers, the frequency emitted in a quantum jump matched, in the limit, the classical orbital rotation frequency. A further example of the parallel between classical and quantum theories is evident if we look at Eqn. 5.15, which gives the classical orbital frequency,

$$f_n = \frac{4\pi^2 m e^4}{n^3 h^3} = \frac{2R}{n^3},$$ (5.15) (6.5)

for an orbit associated with quantum number n ($z = 1$ for hydrogen, R = Rydberg constant). If we integrate over a frequency range from f_n to $f_{n'}$ we have

$$\int_n^{n'} \frac{2R}{n^3} dn = R\left(\frac{1}{n^2} - \frac{1}{n'^2}\right).$$ (6.6)

The right-hand side gives the emitted frequencies (for hydrogen) according to Bohr's theory (cf. Eqn. 5.1) for the $n \to n'$ transition, and the left-hand side amounts to an average of the classical rotation frequencies between the two stationary states. As they become closer, one just has the classical rotation frequency (Eqn. 6.5).

To attempt to find values for the A's and B's, Bohr noted that although a simple linear oscillator has only one frequency, if one considers a non-linear oscillator, then we would have a range of frequencies that can be expressed as a Fourier series of harmonics. In this way, he was able to show a close relationship between the classical frequencies of the Fourier harmonics for such an oscillator and the corresponding frequencies emitted according to the quantum theory. Thus, in Eqn. 6.6 when there is a jump of one quantum number ($n' = n + 1$), the quantum frequency is the average classical frequency of the two orbits, and a jump of two units gives a frequency double that, and so on, i.e. the harmonics of a Fourier series. In general terms, one could say that in the transition $n \to n - \alpha$ in the quantum theory, there corresponds the αth harmonic of the classical motion of the electron in the nth state.

Bohr then assumed that the intensities emitted according to the quantum theory in a given transition would be closely related to the amplitude of the corresponding harmonic.

However, with a non-linear oscillator, the comparison with quantum theory is more complicated than with a linear oscillator and

Bohr's formulae were only accurate for large quantum numbers. Its importance here is that it was the starting point from which Heisenberg and Born, and Dirac developed the important subject of **matrix mechanics** which we deal with in Chap. 8. Before that, in the next chapter, we look at the other important — and easier to visualise — development that was taking place, namely the development of **wave mechanics**. The two forms of mechanics, each with its special virtues, were subsequently shown to be equivalent (§8.3).

Chapter 7

Wave Mechanics

7.1. Introduction

Einstein's 1905 concept of light travelling as **'light quanta'** as a way of understanding details of the photoelectric effect was at odds with all the evidence from diffraction and interference effects that light is a transverse wavemotion. Then, in 1923 came incontrovertible evidence of **wave–particle duality** of X-rays which provided the next step forward.

X-rays, discovered in 1895 by Röntgen (for which he was awarded the first Nobel Prize for physics, in 1901), had been shown to have the same wave properties as visible light — firstly by their polarisation (Barkla, 1906)[c], and secondly by their diffraction (the **Laue Experiment** in 1912 for which Max von Laue was awarded the Nobel Prize for physics in 1914), and the work of W. H. and W. L. Bragg (jointly awarded the 1915 Nobel Prize). However, in 1923, Compton reported the observation that X-rays suffer a wavelength change when they are scattered by the loosely-bound outer orbital electrons in atoms (§4.2). Compton showed that the wavelength change is accounted for quantitatively by considering the scattering process in terms of the conservation of momentum between colliding particles, namely X-ray 'photons' and orbital electrons. The **'Compton effect'** clearly

---

[c] Barkla was awarded the 1917 Nobel Prize for physics for his work identifying the characteristic X-radiation (which he named the K and L lines) of the elements.

reinforced Einstein's theory about the mechanism involved in the photoelectric effect.

A key advance concerning duality was then its extension to being applicable to matter as well as radiation — soon confirmed experimentally. This was the work of de Broglie in 1923, which through subsequent, virtually simultaneous, independent work published by Heisenberg in 1925 and Schrödinger in 1926 heralded — from the different approaches of **matrix mechanics** and **wave mechanics** — the birth of what became known as the '**New Quantum Theory**'.

Bohr's '**Old Quantum Theory**' had played a vital part in the early evolution of the quantum theory but his model of the atom had failed to explain the finer details of atomic spectra and did not explain the mechanical principles involved in causing electrons to reside in specific energy levels or how they jump between levels. Furthermore, it would not have been extendible to deal with duality. However, combined as ever with Einstein's work, the 'Old Quantum Theory' had provided just the right context for the next step forward.

7.2. de Broglie's Matter Waves

In 1923, the year of the Compton Effect, de Broglie published three short papers ('Notes') in the French journal *Comptes Rendus* concerning the nature of light quanta, with a summary published in the following year. It was to be a major advance in physics that was not widely appreciated until his 1924 doctorate thesis on this work at the Sorbonne was published in *Annales de Physique* a year later. We refer to its wide relevance and foresight in §7.3.1, where we note its rôle as an inspiration for Schrödinger's further development of wave mechanics.

In the first of the *Comptes Rendus* papers, de Broglie suggested that nature is symmetrical and that Einstein's wave–particle duality of photons (light quanta) also applies to matter particles. In 1925, Einstein independently came to the same conclusion in his final paper using the **Fluctuation Theory** (§4.4).

De Broglie regarded Einstein's equation $E = h\nu$ for photons as suggesting the "impossibility of considering an isolated fragment of energy without assigning a certain frequency to it". Also, Einstein's expression for **photon momentum** $p = h/\lambda$ (§4.2) would give $\lambda = h/mv$ if the momentum p referred instead to a particle of mass m and speed v. Applied to matter, the problem was the physical meaning of the frequency. He noted that the below equations,

$$E = hv, \quad p = \frac{h}{\lambda},$$

have energy and momentum on the left, associated with particles, and frequency and wavelength on the right, associated with waves.

He also noted that according to Bohr, electrons in atomic orbits occupy distinct energy levels defined by integers, and integers are a characteristic commonly associated with interference effects in optics and with normal modes of vibration in wavemotion.

With such thoughts in mind, de Broglie stated that his starting point was "the wish always to associate the idea of a particle with that of periodicity, in such a way as to bind inseparably the idea of the motion of a particle with that of wave propagation".

De Broglie first examined the simplest case, namely that of a particle "moving freely outside any field of force" and showed that the connection between a particle and associated wave is "in some measure imposed by the fundamental principles of relativity". Another aspect he investigated of relevance here was how the particle–wave concept fitted into Bohr's quantum model of orbital electrons.

7.2.1. *Freely moving particles*

De Broglie assigned a frequency, say v_0, to a simple harmonic wave associated with the particle. Let the wave have the form

$$a \cos 2\pi v_0 t_0,$$

where a = amplitude, and t_0 = time, measured in the coordinate frame of the particle.

Consider an observer in a coordinate system relative to which the particle has velocity v. (For simplicity and without losing generality, we can use one dimension.)

To de Broglie, the essential point was to determine the form of the particle's **associated wave** in the observer's frame of reference. In the latter, the wave will have the form,

$$a \cos 2\pi\nu_0 \left(\frac{t - \upsilon x/c^2}{\sqrt{1 - \upsilon^2/c^2}} \right)$$

$$= a \cos \frac{2\pi\nu_0}{\sqrt{1 - \upsilon^2/c^2}}(t - \upsilon x/c^2), \qquad (7.1)$$

(using the appropriate Lorentz Transformation, cf. §2.3).

Thus, the wave frequency, ν, as seen by the observer is given by

$$\nu = \frac{\nu_0}{\sqrt{1 - \upsilon^2/c^2}} \qquad (7.2)$$

and the 'phase wave' velocity u is given by

$$u = c^2/\upsilon. \qquad (7.3)$$

Clearly the wave does not travel with a speed greater than the speed of light. It suggested dispersion and de Broglie went on to show that it is the group velocity of the wave that matches the velocity of the particle. To do this, he used the phase velocity to "define an index of refraction", n, by what he called 'the usual relation' viz.,

$$n = c/u, \qquad (7.4)$$

(though we would normally write $n(\nu)$ because n is a continuous variable).

De Broglie was therefore intuitively exploring an analogy with light, although the physical significance of n as a refractive index would not be clear (and it would be less than unity).

With Eqn. 7.3 the above gave

$$n = \upsilon/c, \qquad (7.5)$$

and with Eqn. 7.2 he obtained

$$n = \sqrt{1 - \frac{\nu_0^2}{\nu^2}}, \qquad (7.6)$$

indicating **dispersion** of the wavemotion associated with a particle characterised by v_0 .

Group velocity, u_g, is then given by

$$u_g = \frac{dv}{d\left(\frac{1}{\lambda}\right)} \qquad (7.7)$$

(see Appendix G), and to evaluate this we have Eqn. 7.4,

$$\frac{c}{n} = u$$
$$= \lambda v.$$
$$\therefore nv = \frac{c}{\lambda},$$

and Eqn. 7.7 becomes

$$u_g = c\frac{dv}{d(nv)}. \qquad (7.8)$$

From Eqn. 7.6,

$$nv = \sqrt{v^2 - v_0^2}.$$

$$\therefore \frac{d(nv)}{dv} = \frac{v}{\sqrt{v^2 - v_0^2}}$$

$$= \frac{1}{n} \quad \text{by Eqn. 7.6} \qquad (7.9)$$

$$u_g = nc$$
$$= v \quad \text{by Eqn. 7.5,} \qquad (7.10)$$

showing that the group velocity of the associated wave is the same as the velocity of the particle (de Broglie, 1930 — see Bibliography).

De Broglie went on to establish his famous equation relating the mechanical quantity, momentum, of a travelling particle to its wave features. He reasoned that one must expect to find energy, E, of a particle, travelling at velocity v relative to an observer, and its associated wave frequency, v, to be given by the Einstein equation,

$$E = hv. \qquad (7.11)$$

Also, we have

$$E = \frac{mc^2}{\sqrt{1 - v^2/c^2}}, \qquad (7.12)$$

which is Eqn. 2.11.

Similarly, the momentum of the particle is given (Eqn. 2.12) by

$$p = \frac{mv}{\sqrt{1 - v^2/c^2}}, \qquad (7.13)$$

where v is the observed velocity of the particle of mass m.

Using Eqns. 7.11 and 7.12 gives

$$p = \frac{Ev}{c^2},$$

which with Eqn. 7.11 for E and Eqn. 7.3 for v gives

$$p = \frac{hv}{u}. \qquad (7.14)$$

But

$$u = \lambda v, \qquad (7.15)$$

where λ is the observed wavelength of the associated wave.

Therefore,

$$\lambda = \frac{h}{p} \qquad (7.16)$$

$$= \frac{h\sqrt{1 - v^2/c^2}}{mv} \quad \text{by Eqn. 7.13.} \qquad (7.17)$$

Equation 7.16 is the famous **de Broglie Equation**.

In the non-relativistic case, when v^2/c^2 is negligible with respect to unity, de Broglie had thus shown that

$$\lambda = \frac{h}{mv}. \qquad (7.18a)$$

λ is known as the **de Broglie wavelength**.

Also, the non-relativistic kinetic energy for a free particle is $E = p^2/2m$ and this, together with $E = h\nu$ and the above equation, gives

$$\nu = \frac{h}{m\lambda^2}, \qquad (7.18b)$$

where, relating ν with λ, it is known as **de Broglie's Dispersion Law**.

At the time, de Broglie speculated that diffraction of electrons by crystals might be observable, as was already familiar with X-rays. Experimental evidence going some way to support de Broglie's proposal had, unbeknown to him, been obtained earlier, though definitive experimental evidence came in 1927 with the work of Davisson and Germer who demonstrated the scattering of electrons by solids, and in 1928 by G. P. Thomson who demonstrated the diffraction of electrons by metal foils. (Ironically, it had been Thomson's father, J. J. Thomson, who in 1897 established the particle nature of electrons.) A few years later, in 1932, Stern and his collaborators showed diffraction effects at the surfaces of crystals using molecular beams of hydrogen and helium, and in §2.1 we mentioned that what is effectively Young's experiment is being repeated with increasingly larger molecules. However, for the macro-objects we observe in everyday life, their mass is such that the associated wavelength is negligibly small and we are completely unaware of any wave property.

7.2.2. *Orbital electrons as waves*

The Bohr model that described electrons in atoms as existing in orbits defined by integers was reminiscent to de Broglie of the normal modes of vibration in wavemotion. It suggested that a periodicity is associated with an electron in its orbit.

Bohr had shown that the angular momentum of an orbital electron in the hydrogen atom is quantised, with orbital radii restricted to values given by Eqn. 4.17 viz.,

$$m\nu r = nh/2\pi, \qquad (7.19)$$

where n is the 'principal quantum number'.

With $p(=mv) = h/\lambda$ (Eqn. 7.16), this gives

$$2\pi r = n\lambda. \tag{7.20}$$

Thus, the permitted Bohr circular orbits have circumferences corresponding to an integral number of de Broglie wavelengths — indeed reminiscent of **standing waves** in classical wave theory, and here they are very satisfactorily compatible with the quantum features of the Bohr model for hydrogen, whereby an electron circles the nucleus only in orbits that meet the requirement in Eqn. 7.20.

This picture of orbital electrons accompanied by what became known as '**pilot waves**' spreading along an orbit was a profoundly significant step in understanding the atom, since it intimately combined both the particle and wave concepts. However, although groundbreaking in concept, de Broglie knew that his theory was neither clearly nor fully established in detail. It left a number of questions unanswered including, most importantly, the problem of accurately predicting energy levels.

$$* \quad * \quad *$$

De Broglie had initially assumed that his waves were the only reality and that the semblance of a particle was due to some form (he considered several possibilities) of localisation of the waves. But this had to be rejected because it led to conflict with experiments. Then, as we have seen above, he proposed that a '**pilot wave**' accompanies a particle but has an independent physical existence from it. Although the wave would not determine the trajectory of a particle, he later argued at the 1927 Solvay Congress[d] that the probability of a particle being situated at a given point was proportional to the intensity of the wave at that point — as in Born's later 'probability' interpretation of Schrödinger's wave mechanics. However, de Broglie wrote that to use his model to deal with a system of particles was not really practicable: one would have the formidable problem of considering the motion of

---

[d] The élite Solvay Congresses/Conferences in Brussels were attended by leading scientists of the day. They were founded in 1911 by the Belgian chemist/industrialist, Ernest Solvay, and were the scene of legendary arguments, especially between Einstein and Bohr in 1927 and 1930 (Chap. 9).

each particle and its associated wave under the action of fields created by other particles and any external fields.

Despite its shortcomings, the idea that de Broglie had introduced was of profound importance. Although wave–particle dualism had already been established, we have seen earlier that it was a dualism restricted to radiation. In de Broglie's theory, dualism also applies to matter, and that suggested that a profound, general principle in nature was involved. Much of his work anticipated what was to follow in the development of wave mechanics by Schrödinger and we refer to it again in the next section. He was awarded the 1929 Nobel Prize for physics for his discovery of the wave nature of electrons.

Within a few years of de Broglie's work, Schrödinger in 1926 produced a different form of **wave mechanics** which, further developed by Max Born, Dirac, and others, made possible a wide range of successful applications in atomic and nuclear problems. A more satisfactory interpretation of de Broglie's '**pilot waves**' also became clear, as we shall see.

At virtually the same time as the arrival of Schrödinger's theory, Heisenberg published a completely different approach, to become known as **matrix mechanics**. Despite the apparently huge difference between the two theories, a formal mathematical correspondence between them was soon found.

One of the many significant results of all this was that quantum numbers ceased to be assumptions and were revealed as the natural outcome of the new approaches.

7.3. Schrödinger's Wave Mechanics

7.3.1. *Introduction*

We have seen that after Einstein had established that the photoelectric effect is explained by regarding light as having particle-like properties, de Broglie took the bold step of suggesting that wave properties are also associated with moving particles of matter (§7.2) and confirmed experimentally. In de Broglie's model, the wave associated with a moving particle was visualised as a '**pilot wave**'. For orbital electrons in atoms, the permitted orbitals had circumferences corresponding to an integral number of de Broglie wavelengths — reminiscent of standing waves in a stretched string.

However, de Broglie realised that the actual physical nature of his pilot waves was not clear. The model had serious shortcomings, e.g. it provided no way of accurately calculating the energy levels of atomic orbitals. Nevertheless, his work contributed much to what was to follow, and it has been described as a 'treasure trove' (Duck and Sudarshan, 2000). He derived the **Bohr–Sommerfeld quantum condition** $\oint pdq = nh$ (see Appendix C), gave an explanation of interference and coherence properties of quanta, and derived the Compton scattering by moving electrons. Very importantly, he anticipated the work of Bose and Einstein on thermal **radiation as a gas of light quanta**, and also recognised the relevance of Hamilton's theory concerning the similarity between light rays and material particle trajectories. Both the latter topics were taken up in more detail by Schrödinger, who developed a form of **wave mechanics** that was a great advance and had far-reaching significance and usefulness which continues today.

In contrast to de Broglie's travelling waves that accompany moving particles, the waves of Schrödinger's theory can be described as three-dimensional vibration modes (**'eigen-oscillations'**), more akin to the standing waves in acoustics. The **'wave equation'** he developed could be applied to a wide range of examples. Thus, for an atom of a given element, the solution gives the discrete wave patterns for that atom, and with each of those states it gives not only a value for the energy but also yields information about transitions between states. Moreover, the quantum aspects appear naturally instead of having to be artificially introduced as in Bohr's theory.

Schrödinger initially believed his waves to be the fundamental constituents of matter. He saw them as **'wavepackets'** of de Broglie waves, with such packets comprising particles. However, this aspect became untenable.

The theory made its appearance in a four-part paper published in 1926 in the journal *Annalen der Physik*, and in book form a year later.

The actual way in which the wave equation was set up was, as mentioned above, inspired by the work of William Hamilton, the

Irish mathematician and astronomer who in the 1880s had brought together Fermat's 17th century 'Principle of Least Time' concerning the passage of light rays, and Maupertuis' 18th century 'Principle of Least Action' concerning the trajectories of material bodies. More credit should be given to the fact that the relevance of this had already been noted by de Broglie, though he did not pursue it so fully.

So much, for the moment, about the way in which Schrödinger set up his wave mechanics. The philosophical *reasoning* that led him to it stemmed from the work he had just completed and published under the title 'On Einstein's gas theory' (Schrödinger, 1926(d)). He referred to it in the first of the four-part paper referred to above, where he stated "I have recently shown that one can base Einstein's theory of a gas on the idea of standing waves which obey the dispersion law of de Broglie" (Eqn. 7.18b). It was this that led him to use a model of standing waves to also represent particles of matter.

In the next section we deal, very briefly, with the **gas theory** that influenced Schrödinger in taking the direction he chose. We shall then remind ourselves of the classical reconciliation of wave optics with geometrical optics (§7.3.3), and of the early ideas of Fermat and Maupertuis that led to **Hamiltonian mechanics** (§7.3.4–7.3.6). Schrödinger's way of formulating his wave mechanics, with all this as a background, then follows in §7.3.7–7.3.10.

7.3.2. *Background rôle of Einstein's gas theory*

In Chap. 3 we have seen that it was not possible to obtain Planck's highly successful radiation formula from classical statistical methods. Planck had to postulate that the emission of radiation by the resonators, visualised as comprising the walls of the radiation enclosure, is discontinuous. However, he still had to complete the derivation of the radiation distribution function using classical methods. Such a mixed approach clearly suggested that something was wrong.

Then, Einstein's work of 1905 showed radiation itself to have a corpuscular property with quantised energy (§4.3). In that famous March 1905 paper he arrived at the light-quantum postulate from an analogy between radiation and a classical gas of material particles. For the former, he used Wien's **Distribution Law** (Eqn. 3.8) which was based on the concept of gases radiating under thermal excitation. For the latter, he used the expression (Eqn. 4.4), well known in classical thermodynamics, for the volume-dependence of entropy in gases. As we saw earlier (§4.3), it gave a very successful explanation of the **photoelectric effect**. The radiation in a black-body enclosure could, with some justification, be regarded as a '**photon gas**'. Radiation studies could now shift away from thinking in terms of resonators in black-body radiation cavity walls.

During 1909–1917, Einstein made further major contributions to the study of radiation, in an investigation of fluctuations around thermal equilibria. This approach also suggested a wave–particle duality for radiation (see §4.4.1, Eqn. 4.31).

With de Broglie's ideas about matter waves, it is not surprising that Einstein applied his interest in fluctuations to the study of gases. He showed that Eqn. 4.31 could be reformulated to apply to a gas and that, again, one could identify both particle and wave interpretations. He published three papers, one in 1924 and two in 1925, on this and other matters. They were his final contributions in this area.

At this time, Satyendra Bose, in the University of Dacca, sent a paper to Einstein entitled 'Planck's law and the hypothesis of light quanta'. It showed how Planck's radiation law could be derived in a much needed, radically different way. He did this in two steps, each breaking new ground. Firstly, he obtained the factor $8\pi v^2/c^3$ in Planck's formula without any recourse to classical electrodynamics, which had been the basis, in different ways, of the derivations by Planck (§3.2.2) and by Debye (who followed the method of Rayleigh) (see Appendices D and E). Instead, he used Einstein's hypothesis of light quanta (which had been already established), and considered the quanta as a 'photon gas' of 'particles' in **phase space** (see Appendix C), and found that the number of **phase space cells**, each of volume h^3, available to quanta for a given frequency was the same as the factor in Planck's formula.

Next, he had to determine the number of ways in which the quanta could be distributed over the given number of phase cells. There was no constraint on the number of quanta: in exchanges between photons and enclosure walls photons would be absorbed and replaced by the emission of one or more photons of different frequencies — but with the overall energy unchanged. (What Boltzmann would regard as different **microstates** were to be regarded as a single microstate.)

For this, Bose used a different approach to the statistics, in which photons are treated as indistinguishable (unlike Boltzmann statistics), made possible by focusing attention on black-body radiation as a photon gas. Having calculated the number of ways in which the phase-points of quanta could be distributed among the available cells, he then maximised that to give the entropy of the radiation via $S = k \log W$. Planck's radiation formula followed in the familiar way (§3.2.3).

The above description is an oversimplification because we would otherwise need to dig deeply into statistics and probability theory and lose the thread of this story. There had been, and had continued to be, many disagreements about the types of statistics used in this whole area of the subject (e.g. the different statistics used by Planck and Boltzmann) with close relevance to theoretical work at that time on entropy. Even Bose is alleged by Pais to have remarked that he did not know that he was doing something different from Boltzmann's statistics! However, there certainly were new features: no 'particle' conservation, no particle *distinguishability* (a feature of Boltzmann statistics), and no statistical independence of particles.

Other, less complete, contributions along the same lines had been made by others to different extents, notably by Ehrenfest as early as 1911, and de Broglie in 1922.

Einstein was very impressed by Bose's novel and very direct approach to treating radiation. He realised that the method Bose had used was applicable to material gases, and he published three papers (referred to earlier) during 1924–1925 on various aspects. Broadly, they were concerned with the general area of topics such as statistics, absolute entropy, and Nernst's heat theorem. Of just his second paper, on the now-called '**Bose–Einstein gas theory**', Klein (1964) said that it contained "as many ideas in its dozen pages as an annual volume of the journals of physics".

Importantly for us here, Einstein's gas theory added strong support to de Broglie's ideas abut the wave–particle duality of matter. It was this that drew Schrödinger's interest to de Broglie's matter waves, and he was led to conclude that one could use a wave picture of a gas and impose quantisation on that in the way Debye had done for radiation (see Appendix E).

Schrödinger's 1926 paper 'On Einstein's gas theory' was followed by the four-part paper referred to in §7.3.1, detailing the formulation of his wave mechanics to which we turn in §7.3.7, after the preliminaries in the following sections. Before that, it is interesting to summarise the above sequence of developments as follows:

Einstein 1905   Wien's law   + gas as analogue of radiation »» light quanta (photons)

Bose   1924   Photons   + Bose's 'quantum statistics'   »» Planck's law

Einstein 1924–5 Bose statistics + photons as analogue of gas   »» the 'quantum gas'.

7.3.3. *Classical wave optics and geometrical optics compared: a note*

A simple one-dimensional sinusoidal wave can be used successfully as the basis for dealing with many aspects of wave optics. If such a wave has wavelength λ and period T, then the 'wave velocity', (or 'phase velocity'), u, is given by

$$u = \frac{\lambda}{T} = \frac{\nu}{\nu'} \tag{7.21}$$

where frequency $\nu = 1/T$, and **wave number**, $\nu' = 1/\lambda$.

In the **Young–Fresnel theory**, white light may be described as a superposition of sinusoidal waves of various wavelengths and amplitudes. The length of the '**wavetrain**' is limited, however, because the distance over which the component waves are reasonably in step and adding up is limited. Away from that region, the waves become progressively out of step and cancel out by interference. The resulting 'wave group' travels like a single entity with '**group velocity**', u_g, given by

$$u_g = \frac{d\nu}{d\nu'}, \tag{7.22}$$

(see Appendix G, Eqn. G.5). Note that only if there is no **dispersion**, the phase velocity is the same as the group velocity. In general, the wavemotion moves through the group as it advances.

When the size of the wave groups is comparable with the dimensions of the experimental apparatus (e.g. its apertures and the distances between them), we observe diffraction and interference effects as demonstrated originally by Young in 1801 in his double-aperture experiment (Fig. 2.1). This is the realm of physical optics. In contrast, the laws of geometrical optics apply when the apparatus dimensions are very large compared with the lengths of the wave groups. The wave groups are then practically points: interference and diffraction effects cannot occur and the path followed is what we traditionally call a 'ray' (Fig. 7.1).

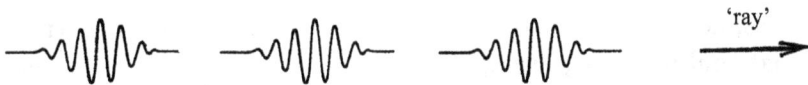

Figure 7.1. Exaggerated depiction of wave groups comprising a 'ray' in the geometrical optics sense.

7.3.4. *Fermat's Principle of Least Time*

Pierre de Fermat (1601–1665) reasoned that the laws of geometrical optics are based on the single requirement that a light ray passing from point A to point B takes a path such that the number of wavelengths sweeping through the path has a value that is 'stationary' with *respect to adjacent alternatives*. Note that this refers to the optical path, i.e. with due allowance for the changes in wavelength if the ray passes through regions of different refractive index. It can be expressed as

$$\delta \int_{A}^{B} v'.dl = 0, \tag{7.23}$$

where v' is **wave number** $(1/\lambda)$ and dl is an element of length along the path A to B . δ represents the differential with respect to the lengths of neighbouring, possibly, alternative paths and thereby seeks the stationary value of the integral.

Note that v' may vary along the path A to B since the refractive index may not be the same throughout (whereas the frequency is constant).

Note also that Fermat's Principle as expressed above only decides the 'shape' of the ray path. Although it is traditionally known as the **Principle of Least Time**, that name is misleading. If, by 'time' it refers to the time taken by e.g. a wave crest to travel from A to B then it will of course also have a stationary value because it is proportional to the number of waves in the path. However, the experimentally observed time is the time taken for the wave group to travel the path and the times are not in general the same.

[It has been pointed out by various authors that 2,000 years or so ago, Hero of Alexandria asserted that the reflection of light by a mirror follows the shortest path between an object and its reflection. In this, he was the first to express and use what became known as a **variational principle**, exemplified by Eqn. 7.23. Laws expressible in this form are of course often expressed in other ways. For example, the laws of refraction and reflection as we usually use them are expressed in a non-minimalising form.]

In the next section, we are going to compare Fermat's Principle in optics with **Maupertuis' Principle of Least Action** in mechanics as applied to the trajectories of matter particles, where stationarity will refer to possible adjacent paths being co-terminus in time as well as space. The trajectories of matter particles are analogous to the path of light wave-groups but Fermat's Principle deals, as noted above, only with paths that are co-terminus in space. For a comparison with Maupertuis' Principle, the requirement that an optical path is also stationary with respect to time can be expressed as

$$\delta \int_A^B v.dt = 0, \qquad (7.24)$$

where $v.dt$ is the product of frequency (unlike v', this is constant regardless of path) and time. For the comparison, the two requirements are combined as an **Extended Fermat's Principle:**

$$\delta \int_A^B (v'.dl - v.dt) = 0. \qquad (7.25)$$

7.3.5. *Maupertuis' Principle of Least Action*

In c. 1745, a century after Fermat enunciated his principle, Pierre–Louis Moreau de Maupertuis (1698–1759) arrived at his '**Principle of Least Action**' in mechanics. He developed this from what his biographer H. B. Glass (1955) described as "a feeling that the perfection of the universe demands a certain economy in nature and is opposed to any needless expenditure of energy. Natural motions must be such as to make some quantity a minimum". Maupertuis concluded that in a dynamic system changing conservatively[e] from state A to state B, the quantity that tends to a minimum is the product of what today we call the kinetic energy, $T\,(= \frac{1}{2}mv^2)$, and the duration of the movement. The product was named the '**action**' [units of energy × time, a four-dimensional equivalent of energy (§3.2.3)], and the requirement that it is a minimum was originally expressed by Maupertuis as

$$\delta \int_A^B 2T.dt = 0,$$

which can alternatively be expressed as $\qquad\qquad\qquad$ (7.26)

$$\delta \int_A^B p.dq = 0,$$

where p denotes the momentum as earlier in this chapter, and q denotes the spatial coordinate.

Note that '**action**' is therefore also seen to be expressed as the product of momentum and distance.

The above should strictly be called the **Principle of Stationary Action** since the action is not in all cases a minimum though in our applications here it will be.

Maupertuis regarded his principle in more general terms to be applicable to all natural phenomena. Today it can be seen to be basic to many aspects of the animate as well as the inanimate world but

---

[e] It is important to note that the principle applies to a conservative system i.e. the total internal energy of the system in going from A to B is conserved and any internal work performed is determined by the terminal configurations independently of the particular sequence of intermediate configuration.

at the time, his view of its universal applicability elicited adverse criticism — Jeans regarded it as theological and metaphysical rather than scientific. Maupertuis became a forgotten genius, and the more explicit statement of his principle applied to dynamics that we know today (Eqn. 7.26) is often attributed variously to Euler, Lagrange, and Hamilton. To offset his neglect, Glass notes that he was the first to apply the laws of probability to the study of heredity, that he fore-cast genetic theory, and that his various studies of evolution rank him above all other precursors of Darwin. At one time, he was president of the Academy of Sciences in Berlin, and also headed an expedition to Lapland, confirming the flattening of the Earth towards the poles in accordance with Newton's gravitation theory. This was no mean achievement.

In the original form in which Maupertuis stated his Principle (Eqn. 7.26), the requirement for minimum action concerned paths that are co-terminus in space but not in time. Just as we have included time in Fermat's Principle in Eqn. 7.25, it is shown in **Classical Mechanics** that Maupertuis' original statement (Eqn. 7.26) then becomes

$$\delta \int_A^B (T - V)\, dt = 0, \qquad (7.27)$$

where V is potential energy.

A more instructive version is obtained by writing the above as

$$\delta \int_A^B (2T - E)\, dt = 0, \qquad (7.28)$$

where $E = T + V$, and hence

$$\delta \int_A^B \left(p\frac{dq}{dt} - E \right) dt = 0, \qquad (7.29)$$

and therefore

$$\delta \int_A^B (p\,dq - E\,dt) = 0, \qquad (7.30a)$$

where p and q are conventionally the momentum and distance respec-tively (we return to their wider use in Chap. 8).

It is the comparison of this with the form of Fermat's Principle in Eqn. 7.25 that is so interesting and to which we return in the next section.

The integral in Eqn. 7.30a is often referred to as the '**action integral**'. It provides a sufficient condition for deriving equations of motion rather than Newton's laws of motion. Furthermore, it has the advantage that the integral is invariant to the system of generalised coordinates. This so-called '**variational principle**' enables non-mechanical systems to be dealt with in the same way — as we have seen with Fermat's Principle.

7.3.6. *Hamilton's mechanics and Schrödinger's wave mechanics*

The general subject of **Classical Mechanics** is concerned with the mathematical developments that were introduced for dealing with the use and application of the basic tenets of Newton's laws of motion. It was the particular form of classical mechanics that Hamilton developed that profoundly influenced not only Schrödinger but also the later development of quantum mechanics. Hamilton's mechanics stemmed from his study of problems in geometrical optics in which he used the analogy between geometrical optics and mechanics. He had been struck by the way in which Fermat's principle in optics (§7.3.3) and Maupertuis' principle in dynamics (§7.3.4) both involve stationary values in determining the correct paths taken by light rays in one case and the trajectories of material bodies in the other. If, in Fermat's Principle as expressed in Eqn. 7.25, we replace $v'(= 1/\lambda)$ by p/h (Eqn. 7.16) and v by $E = hv$ (Eqn. 7.11), we get Eqn. 7.30a, which is effectively **Hamilton's Principle**. The full analogy is as follows:

Geometrical optical path Fermat		Mechanical trajectory Maupertuis
wave number v'	- - - - -	momentum p
distance l	- - - - -	distance q
frequency v	- - - - -	energy E
v	- - - - - conserved - - - - -	E

Hamilton's Principle is usually written with H instead of E as a recognition of his work:

$$\delta \int_A^B (pdq - Hdt) = 0. \qquad (7.30b)$$

An obvious inference from the analogy was that any purely mechanical problem (i.e. not involving heat) can be translated into a problem in optics in the limiting case where geometrical optics applies. Thus momentum can replace wave number and energy can replace frequency, though of course there is no physical reality in this.

Influenced very much by the analogy, Hamilton developed a branch of mechanics which included his own 'wave mechanics' though the latter aspect was regarded at the time as a mathematical curiosity. In his mechanics, the state of a conservative system with n degrees of freedom is determined at any instant by coordinates q specifying the configuration, and p specifying the corresponding momenta. The total energy is given by an 'appropriate combination' of all the (n) p's and q's, and denoted by $H(q, p)$. This **Hamiltonian function**' is readily constructed for any dynamic system, especially if the coordinates can be chosen so that H is only a function of the corresponding p's. In this way, Hamilton developed the '**canonical equations**' in mechanics that are named after him. ['Canonical' here means nothing more than that they are a set of standard equations in Hamilton's mechanics, and p and q are referred to as '**canonical variables**' in that set of equations. Goldstein states that the term was probably first introduced by C. G. J. Jacobi in 1837 in an application of Hamilton's equations, and he adds that others later wondered why the term was necessary.]

The relevance of Hamilton's mechanics for us here was the conclusion drawn when Fermat's and Maupertuis' principles break down. In an optical system when the dimensions of apertures, optical paths etc. become so small as to be comparable with the wavelength of the light, the laws of geometrical optics are not applicable and, as we know, a wave model is appropriate. This failure of geometrical optics has its parallel in classical mechanics which break down when applied to sufficiently small systems such as atoms, electrons, etc. (as first recognised by Planck). This suggested that with very small systems, it is necessary to change to a particle–wave duality for matter.

Though de Broglie had originated the wave mechanics, it was Schrödinger who contributed greatly to its development as a result of exploring more fully the nature and importance of the Hamilton analogy. He pursued the idea that if geometrical (ray) optics is the large-scale approximation of wave-optics, then classical mechanics is an approximation of a more fundamental mechanics, a new 'wave mechanics' that would show the finer details of the behaviour of matter particles. **Duality** was therefore not to be regarded as unique to the quantum world underlying de Broglie's wave mechanics but should hold at all levels. Thus, we have

Geometrical optics : wave optics = classical mechanics : wave mechanics.

$$(7.31)$$

7.3.7. *Schrödinger's time-independent wave equation of 1926*

In the four-part paper published in 1926, Schrödinger progressively investigated in great detail a number of aspects of quantisation from the point of view he was developing. In embarking on the study, he was concerned that up till then 'whole numbers' entered quantum rules 'mysteriously'. We shall see that in the theory he developed, this 'integralness' would have its origin in the 'finiteness and single-valuedness of a certain space function'.

In a lecture in 1928, he described the essence of what had been his 'derivation of a fundamental idea of wave mechanics from Hamilton's analogy between mechanics and geometrical optics'. Schrödinger used the statement of the principle as given to it by Maupertuis (Eqn. 7.26) namely,

$$\delta \int_A^B 2T.dt = 0, \qquad (7.32)$$

where T is the kinetic energy of a 'mass-point' m in a conservative field of force described by the assumed time-independent potential energy $V(x)$.

If the mass-point particle velocity is v we then have

$$T = \tfrac{1}{2}mv^2$$
$$= E - V, \tag{7.33}$$

where E is the total energy of the particle. This is the **Classical Energy Equation.**

Since $v = ds/dt$, using s again as the distance variable, Eqn. 7.32 can be written as

$$\delta \int_A^B \sqrt{2m(E-V)}\,ds = 0. \tag{7.34}$$

Hamilton had compared this with Fermat's Principle in optics, as expressed in Eqn. 7.24, the two becoming identical if we put phase velocity,

$$u = \frac{C}{\sqrt{2m(E-V)}}, \tag{7.35}$$

where C is a constant.

This implies that one can visualise the ray path followed by light rays in an optical system as also being the trajectory of a 'mass-point' m moving with given energy E under the influence of a 'field of force' $V(x)$. It is a sign of Schrödinger's genius that he regarded as a matter of great importance the fact that in the above equation the velocity of light depends not only on coordinates along a ray path but also on E, the total energy of the corresponding mass point in its trajectory. He felt that this enabled him to "push the analogy a step further by picturing the dependence of u on E as dispersion, i.e. a dependence on frequency". To the path of light rays that would correspond to the trajectory of a material particle, he therefore attributed a frequency v given ('arbitrarily') by the Einstein equation $E = hv$. The next step was to see if he could "make a small 'point-like' signal move exactly like our mass point". He used the word 'exactly' because, as he went on to add, hitherto there had only been geometrical identity of the trajectories and a neglect of the question of speeds. A light signal moves with a group velocity and it was therefore necessary to see if this would match the velocity of a mass point which, from Eqn. 7.33, is given by

$$v = \frac{\sqrt{2m\,(E - V)}}{m}.$$
(7.36)

Now the group velocity, u_g, of the light signal is given as usual (see Appendix G) by

$$u_g = \frac{dv}{d\,(1/\lambda)}$$

$$= \frac{dv}{d\,(v/u)},$$
(7.37)

where u = phase velocity.

With Einstein's $E = hv$, we therefore have

$$u_g = \frac{dE}{d\,(E/u)}.$$
(7.38)

Substituting from Eqn. 7.35 for u gives

$$u_g = \frac{dE}{d\left[E\sqrt{2m\,(E - V)}/C\right]}.$$
(7.39)

To compare this with the velocity v (Eqn. 7.36) of the mass particle, we can express v as

$$v = \frac{dE}{d\left(\sqrt{2m\,(E - V)}\right)}$$
(7.40)

$$\left(\frac{d\sqrt{ax + b}}{dx} = \frac{1}{2}\frac{a}{\sqrt{ax + b}}\right)$$

Thus, if $C = E$, there is coincidence, as Schrödinger put it, 'between the dynamical laws of motion of the mass-point and the optical laws of motion of light-signals in our imagined light-propagation'.

Equation 7.35 for the phase velocity becomes

$$u = \frac{E}{\sqrt{2m\,(E - V)}},$$
(7.41)

and with $u = \lambda v$ and $E = hv$, we obtain

$$\lambda = \frac{u}{v} = \frac{uh}{E} = \frac{h}{\sqrt{2m(E-V)}}. \tag{7.42}$$

With Eqn. 7.36, this gives

$$\lambda = \frac{h}{mv}. \tag{7.43}$$

This is in agreement with de Broglie's equation (Eqn. 7.18), but in contrast to his derivation, we see that it was arrived at by Schrödinger purely as a consequence of Hamilton's analogy and Einstein's $E = hv$.

Here was the implication that wave–particle duality is not unique to quantum matters, and Schrödinger reasoned that the mechanics-optics parallel should hold at all levels. He therefore proceeded to adopt the basic equation of wave optics for application to matter. We can write the equation here in the classical one-dimensional form for simplicity as

$$\frac{\partial^2 \Psi(x,t)}{\partial x^2} - \frac{1}{u^2} \cdot \frac{\partial^2 \Psi(x,t)}{\partial t^2} = 0, \tag{7.44}$$

where Ψ (a function of x and t) is the wave amplitude and u is the phase velocity (Eqn. 7.41), giving

$$\frac{\partial^2 \Psi(x,t)}{\partial x^2} - \frac{2m(E-V)}{E^2} \cdot \frac{\partial^2 \Psi(x,t)}{\partial t^2} = 0. \tag{7.45}$$

Although time-dependent, this assumes that V itself is time-independent and therefore this equation applies only to 'conservative systems' for which the total energy $E(=T+V)$ is constant. Furthermore, general solutions of this equation would lead to negative as well as positive energies which, for a free particle, would mean negative kinetic energy. Schrödinger could not, therefore, use Eqn. 7.45 as a time-independent wave equation. Writing the general solution to Eqn. 7.45 as

$$\frac{\partial^2 \Psi(x,t)}{\partial t^2} = -\omega^2 \Psi(x,t), \tag{7.46}$$

Eqn. 7.45 becomes

$$\frac{\partial^2 \Psi(x,t)}{\partial x^2} + \frac{2m(E-V)}{E^2} \omega^2 \Psi(x,t) = 0. \tag{7.47}$$

For time-independence, we can now use just the $\Psi(x)$ part from Eqn. 7.46 and write

$$\frac{\partial^2 \Psi(x)}{\partial x^2} + \frac{2m(E-V)}{E^2}\omega^2 \Psi(x) = 0. \qquad (7.48)$$

With $\omega = 2\pi\nu$ and using $E = h\nu$ gives $\omega = 2\pi E/h$, and the above can be written as

$$\frac{\partial^2 \Psi}{\partial x^2} + \frac{8\pi^2 m}{h^2}(E-V)\Psi = 0, \qquad (7.49)$$

which is the much quoted and used expression of the famous **Schrödinger time-independent wave equation**.

[Many elementary textbooks derive the Schrödinger equation starting from de Broglie's equation $p = h/\lambda$. To obtain the phase velocity arrived at by Schrödinger, one then uses

$$u = \lambda\nu$$

and

$$E = h\nu$$

to give

$$u = \frac{\lambda E}{h}.$$

With de Broglie's

$$p = \frac{h}{\lambda},$$

then

$$u = \frac{E}{p},$$

and for a conservative field of potential V as here envisaged, we have the '**classical energy equation**' (Eqn. 7.33):

$$E = \tfrac{1}{2}mv^2 + V$$

$$= \frac{p^2}{2m} + V,$$

i.e.

$$p = \sqrt{2m(E-V)},$$

giving

$$u = \frac{E}{\sqrt{2m(E-V)}},$$

as in Eqn. 7.41 and hence to Eqn. 7.44. However, this completely misses the basis of Schrödinger's approach.]

The way in which Schrödinger's equation can be used becomes apparent when it is applied to any particular set of conditions: the function V, boundary conditions etc. Just as the simple wave equation of a vibrating string yields a discrete set of stationary-state solutions when the ends of the string are fixed, so in general the solutions to the Schrödinger equation give certain energy values of E (though some examples involve a continuous spectrum of energy values). Allowed energy values are variously called proper-, characteristic-, or eigen-values for the system.

An immediate example for us here was Schrödinger's application of the three-dimensional version of Eqn. 7.49 to the hydrogen atom. Using a simple potential energy function $V = -e^2/r$ for an orbital electron (e = electron charge, r = its distance from the nucleus), solutions only exist for discrete energy levels given by

$$E_n = -\frac{2\pi^2 m e^4}{n^2 h^2},$$

and hence

$$E_2 - E_1 = v_{1,2} = \frac{2\pi^2 m e^4}{h^2}\left(\frac{1}{n_1^2} - \frac{1}{n_2^2}\right)$$

(7.50)

as arrived at by Bohr's method (cf. Eqn. 5.16), yet without — and this is the important point — any recourse to Bohr's arbitrary postulates about quantum jumps from one stationary state to another. For a given function V, the solutions to the wave equation give the allowed energy levels directly.[f] [de Broglie's treatment (§7.2), also, had been unsatisfactory because it involved mysterious waves accompanying

---

[f] The calculation that gives Eqn. 7.50 is three-dimensional and is best done by changing to spherical coordinates as shown in textbooks dealing with the applications of Schrödinger's equation.

electrons, the wavelengths of the waves being determined by the mass and velocity of the electrons. Permitted orbits were shown to have circumferences corresponding to an integral number of de Broglie wavelengths, but the method was very limited as we noted in §7.2.]

Schrödinger's wave equation is a linear equation. Such equations have the property that when particular solutions are found for a given application, the sum of them is also a solution. The solutions, the **proper** (or **eigen**) **functions**, resemble vectors in **Euclidean** (or **Minkowskian** i.e. four-dimensional) **space** but are represented in a different space known as **Hilbert Space** where 'dimensions' have nothing to do with the physical dimensions of the system to which they refer. [Hilbert Space is named after the German mathematician David Hilbert (1862–1943) who originated the general concept at the beginning of the 20th century and which was applied in 1927 to quantum mechanics by the Hungarian-born US mathematician John von Neumann (1903–1957).]

Despite the mystery of the meaning of the wave equation at that time, we have noted earlier the successful application of it to hydrogen spectra.

The equation, and an extension of it to a time-dependent version (§7.3.8), have been extremely successful in many kinds of atomic and molecular problems, far beyond what could be achieved with Bohr's theory. For example, it successfully provided an explanation of α-particle emission by radioactive elements, and their penetration into the nuclei of other lighter elements with the resulting transmutation of elements.

We should note that this 'wave mechanics' was established without the assistance of ad hoc devices such as the **Correspondence Principle**. In fact Schrödinger even hoped that the success of his wave equation would remove the concept of quantum jumps and could restore to atomic physics the classical principles of continuity which had been losing ground since Planck's paper of 1900 on black-body radiation. As Max Jammer has stated, the Schrödinger four-part paper of 1926 was "undoubtedly one of the most influential contributions ever made in the history of science ... In fact, the subsequent development of non-relativistic quantum theory was to no small extent merely an elaboration and application of Schrödinger's work".

However, despite the huge success of the Schrödinger equation, the actual meaning of Ψ and how the vital concepts of position and velocity fit in, posed problems. Schrödinger regarded the waves in his theory as the fundamental constituents of matter, although it was not clear precisely in what way. For example, he was certainly not attributing to an electron the features commonly associated with a particle, such as being situated at a definite point in space and moving with a definite speed. Instead, he visualised it as a wave-packet corresponding in some way to the electron charge distribution, where group velocity is related to what would be particle velocity. However, it was not appreciated that a wave-packet would spread spatially with time and could not sustain a role of depicting a particle. We return to this in §7.3.9.

[de Broglie subsequently made the interesting observation that his theory was directly compatible with Fermat's and Maupertuis' principles. When a light ray travels from point A to point B, it follows a path in which the line integral

$$\int_A^B \frac{dl}{\lambda},$$

is a minimum. This is the basis of Fermat's Principle (Eqn. 7.23) and with de Broglie's $1/\lambda = p/h$ (Eqn. 7.16), it can be written as proportional to

$$\int_A^B p.dl,$$

which is an expression of **Maupertuis' principle** for particle trajectories (§7.3.4).]

7.3.8. *Schrödinger's time-dependent equation*

A time-dependent equation must of course be consistent with Einstein's energy equation $E = h\nu$ and, to be applicable to particles, with de Broglie's equation $p = h/\lambda$ for momentum. It has also to be consistent with the **Classical Energy Equation** for total energy (at speeds small compared with that of light):

$$\frac{p^2}{2m} + V = E, \tag{7.51}$$

where $p^2/2m$ = kinetic energy, V = potential energy assumed constant (V_0) for the moment (a 'free' particle).

In terms of wavemotion, the wave function Ψ in wave mechanics is regarded as analogous to the transverse displacement $y(x)$ of wavemotion travelling along a stretched string where, as we know, various solutions are possible according to the particular conditions operating. In that well-known analogy, the wave motion is given by the classical wave equation

$$\frac{\partial^2 y}{\partial x^2} = \frac{1}{v^2}\frac{\partial^2 y}{\partial t^2}, \tag{7.52}$$

for which a solution for a wave-train travelling in the $+x$ direction can be expressed as

$$y = Ae^{-2\pi i(vt - x/\lambda)}. \tag{7.53}$$

For a free particle being described by the wave function Ψ, the requirements of $E = h\nu$ and $p = h/\lambda$ are applied, and the above becomes

$$y = \Psi = Ae^{-(2\pi i/h)(Et - px)}. \tag{7.54}$$

For this to be consistent with Eqn. 7.51, we proceed as follows. Differentiate Eqn. 7.54 with respect to x,

$$\frac{\partial \Psi}{\partial x} = \frac{2\pi i}{h}p\Psi, \tag{7.55}$$

$$\frac{\partial^2 \Psi}{\partial x^2} = -\frac{4\pi^2 p^2}{h^2}\Psi, \tag{7.56}$$

and with respect to time,

$$\frac{\partial \Psi}{\partial t} = -\frac{2\pi i E}{h}\Psi. \tag{7.57}$$

Multiply both sides of Eqn. 7.51 by Ψ,

$$E\Psi = \frac{p^2\Psi}{2m} + V_0\Psi, \tag{7.58}$$

and substitute for p^2 from Eqn. 7.56 and for Ψ from Eqn. 7.55, giving

$$-\frac{h^2}{8\pi^2 m}\frac{\partial^2\Psi}{\partial x^2} + V_0\Psi = \frac{ih}{2\pi}\frac{\partial\Psi}{\partial t}, \tag{7.59}$$

which is Schrödinger's **time-dependent 'free particle' equation** (stated here in one-dimensional form for simplicity).

Since this equation is linear in Ψ, it follows that any linear combination of solutions of the form in Eqn. 7.53 is also a solution. Note that it is only first-order in time, not second-order as for the space variables.

So far, the potential has been taken as constant (V_0). However, any spatially continuously varying potential function can be approximated by a series of infinitesimal steps in each of which potential is constant. It was therefore considered reasonable to regard potential in Eqn. 7.59 as a continuous function of x.

When more complicated force fields were considered, e.g. the electrostatic field holding an electron in an atom, the type of wave function solution in Eqn. 7.54 would be inadequate, so a more general form of the wave equation was postulated, with V_0 now expressed as $V(x,t)$,

$$-\frac{h^2}{8\pi^2 m}\frac{\partial^2\Psi}{\partial x^2} + V(x,t)\,\Psi = \frac{ih}{2\pi}\frac{\partial\Psi}{\partial t}. \tag{7.60}$$

This 'time-dependent Schrödinger equation' has a very broad validity. It was the form given in more detail by Schrödinger in the fourth of his 1926 papers in *Annalen der Physik*.

It greatly pleased Schrödinger that the formulation of his wave mechanics was in the familiar language of differential equations and continuous fields. It did not require quantisation to be artificially postulated, as was necessary in Bohr's quantum theory. Quantisation of energy levels here occurs because of the boundary conditions imposed on the wave functions: the wave function should remain finite even when x tends to infinity. This happens automatically for a free particle because its wave function is a combination of sine and cosine terms which are always between -1 and 1, so its energy is not quantised. However, for potentials which vary with position, the wave functions for bound particles contain terms which increase exponentially as x goes to plus or minus infinity, except when the value of E belongs to a particular set of discrete values. If E is greater than the maximum value of V, then the particle has escaped and is free. Its wave function now oscillates at large distances from the centre of the potential, and there is no need to apply a boundary condition, so E is not quantised when $E > V_{max}$.

Furthermore, there was no requirement that atomic systems 'jump' from one stationary state to another in transitions — a concept quite unacceptable to Schrödinger's philosophy of continuity in nature. Emitted radiation resulting from a transition between two states occurs as a '**beat frequency**' between the wave forms associated with the two states. Not surprisingly, Bohr and Heisenberg [who had just produced his matrix mechanics (Chap. 8)] were opposed to this picture because it contradicted Planck's quantum hypothesis and Einstein's photon hypothesis.

Schrödinger's wave mechanics solved a wide variety of problems. It was in this 1926 work that Schrödinger first saw the possibility of what later became known as '**entanglement** (*Verschrankung*) though he did not introduce the term explicitly until 1935 in the paper in which he discussed Einstein's objections to Bohr's belief in '**complementarity**'. For the present, we can describe entanglement as meaning that if two systems interact temporarily, they never behave again as separate systems no matter how far apart they become separated. We return to this in Chap. 10.

Schrödinger shared the 1933 Nobel Prize for physics with Dirac for their 'discovery of new and productive forms of atomic theory'.

Though a major advance, the physical interpretation of Schrödinger's waves, in which the particles of corpuscular physics were regarded as wave-packets, was soon seen to be unsatisfactory, e.g. they would not be stable, they would dissipate (spread).

However, the interpretation had played its part because within a year, the waves were going to be given a much more acceptable interpretation, the 'probabilistic interpretation', to which we now turn.

7.3.9. *The probabilistic interpretation of the wave equation*

The realisation that wave-packets as particles would not be stable but would dissipate, led to the notion in several minds that, instead, the waves of Schrödinger's wave mechanics indicate the probability of particle location. The establishment of this interpretation is associated mainly with Born (1926). However, it had been envisaged by Einstein in 1925 when he suggested that in the relationship between an electromagnetic wave field and light quanta, the waves are a 'phantom (or ghost) field' that guides the path of the corpuscular light quanta. To link the interpretation with electromagnetic wave theory, Einstein had postulated that the optical wave amplitude is large where photons are likely to be found, and where the amplitude is small the likelihood is low. (Einstein's biographer Abraham Pais commented that Einstein had yet again been a "transitional figure in the period of the birth of quantum mechanics".)

Noting that the scattering of light by, for example, dust particles, had been satisfactorily worked out years before by Rayleigh using the classical wave theory, Born felt that if the idea of light quanta was now to be applied, then the observed light intensity must be proportional to the number of light quanta. This suggested using wave mechanics to calculate the scattering of electrons by atoms, and it indeed gave agreement with Rutherford's experimental scattering work and formula. (see e.g. Born's *Atomic Physics*)

In terms of the Schrödinger wave equation, $|\Psi|^2$ would then indicate the probability of there being a particle at the place for which Ψ had been evaluated. The waves of wave mechanics then became **probability waves**, the 'particles' retaining their status as particles. In what Schrödinger (disapprovingly) called a 'makeshift' way, wave and particle aspects were brought together in the same picture.

In terms of Born's interpretation of Schrödinger's wave function, Ψ, if we write

$$P = \Psi\Psi^* = |\Psi|^2, \qquad (7.61)$$

then Pdx would be the probability that at time t a particle would be found between x and $x + dx$. Since the particle must be somewhere, we have

$$\int_{-\infty}^{+\infty} Pdx = \int_{-\infty}^{+\infty} \Psi\Psi^* dx = 1, \qquad (7.62)$$

which provides 'normalisation'.

To calculate Ψ, the independent variables must include time as well as the coordinates of the particles involved. Furthermore, as noted in §7.3.8, the wave equation is linear in Ψ . Thus, if Ψ_1 and Ψ_2 are solutions, then $\Psi_1 + \Psi_2$ must also be a solution. For example, if the **two-slit experiment** (Fig. 2.1) is performed with electrons, then if Ψ_1 represents the propagation through only one slit alone and Ψ_2 represents the propagation through the other slit alone, then $\Psi_1 + \Psi_2$ gives the result if both slits are open. Whereas with the classical idea of probability where the probabilities would be additive, we would now have the probability given by

$$\begin{aligned}
P_{12} &= (\Psi_1 + \Psi_2)^* (\Psi_1 + \Psi_2) \\
&= \Psi_1^*\Psi_1 + \Psi_2^*\Psi_2 + \Psi_1^*\Psi_2 + \Psi_2^*\Psi_1 \\
&= P_1 + P_2 + \Psi_1^*\Psi_2 + \Psi_2^*\Psi_1.
\end{aligned} \qquad (7.63)$$

The last two terms are the familiar '**interference terms**'. We return to discuss this in more detail in §9.3 in the context of wave–particle duality and the **Copenhagen Interpretation**.

Born applied the concept successfully to various problems. As a result, he stated that he was "inclined to renounce determinism in the atomic world".

However, Born's Theory omitted much detail concerning what happened to the wave function underlying his probabilistic interpretation. For example, his interpretation meant that if an electron, as a particle, is at any instant at a particular place then the uncertainty of where to find it can only refer to the ignorance of the observer. Then if the observer actually finds it in a particular place the probability of its being where it is found (by the observer) is unity, and of being elsewhere it is zero. The latter was subsequently seen as the '**collapse of the wave function**', or '**state-vector reduction**', terms originating with John von Neumann that have permeated (not without criticism) discussions of the subject ever since. It was certainly a concept not welcomed by Schrödinger.

Born's papers had a mixed reception. Schrödinger never abandoned a belief in determinism and the fundamental continuity of nature: it was a classical field interpretation. He apparently told Born that he might not have published his papers had he been able to "foresee what consequences they would unleash."

Although Einstein was enthusiastic about Schrödinger's wave mechanics (with which he had a link via its use of his $E = h\nu$ (cf. Eqs. 6.3 and 6.4)) he was always sceptical about Born's probabilistic interpretation. Despite the fact that in 1916 he had laid the foundations of the statistical rules of quantum theory, he could not accept the implications of a physical world that is fundamentally indeterminate: in response to a letter from Born he stated his conviction usually quoted as "**God does not play dice**". However, Einstein's belief that underlying quantum theory there must be a deeper, deterministic theory of nature, lost favour in the face of the huge success of quantum theory — which caused most physicists to come to believe that the ultimate physical theory is quantum mechanical and that nature is fundamentally random.

Einstein had a long-running 'conflict' (his word) with Bohr over the latter's reluctance to accept the photon concept of light transmission, despite Bohr's introduction of the quantum in explaining the generation of **optical line-spectra**. However, Bohr remained convinced about the reality of the quantum jumps that Schrödinger hoped to have dispensed with.

To add to the confusion, the completely different '**matrix mechanics**' of Heisenberg had almost simultaneously appeared on the scene (Chap. 8).

In Chap. 9, we return to the problem of resolving these conflicting views, and where, among other things, we shall see that the probabilistic interpretation was destined, in a wider context, to become a component of Heisenberg's Uncertainty Principle and the subsequent Copenhagen Interpretation.

Born was awarded the 1954 Nobel Prize for physics for his "fundamental work in quantum mechanics and especially for his statistical interpretation of the wave function". (The 1954 Prize was shared with Walther Bothe for the latter's unrelated work on particle detection.)

7.3.10. *Operator representations*

Turning back to Eqn. 7.55 and using the p, q notation, we can write

$$\hat{p} = -\frac{ih}{2\pi}\frac{\partial}{\partial q}, \tag{7.64}$$

implying that momentum is associated with a '**differential operator**' given by the right side of the equation (the '**hat symbol**' is the convention for denoting an operator). Schrödinger used this when he showed the equivalence of his wave mechanics to Heisenberg's matrix mechanics (§8.3).

In the same way, from Eqn. 7.57, we write

$$\hat{E} = \frac{ih}{2\pi}\frac{\partial}{\partial t}, \tag{7.65}$$

which is the operator associated with energy.

This idea of using operators seems daunting but it is no different from the familiar dy/dx, where the differential operator d/dx is acting on y. They are much used in quantum mechanics where they were introduced by Max Born and Norbert Wiener.

The Classical Energy Equation (Eqn. 7.51) viz.

$$\frac{p^2}{2m} + V = E, \tag{7.66}$$

is widely used in operator form. Noting that the multiplication of operators is performed by applying them in order, one after the other, we obtain from Eqn. 7.64,

$$\hat{p}^2 = -\frac{ih}{2\pi}\frac{\partial}{\partial q}\left(-\frac{ih}{2\pi}\frac{\partial}{\partial q}\right)$$

$$= -\frac{h^2}{4\pi^2}\frac{\partial^2}{\partial q^2}. \tag{7.67}$$

With this, and $\hat{E}$ given as in Eqn. 7.65, the operator version of Eqn. 7.66 is

$$-\frac{h^2}{8\pi^2 m}\left(\frac{\partial^2}{\partial q^2}\right) + V = \frac{ih}{2\pi}\frac{\partial}{\partial t}. \tag{7.68}$$

Applied to a wave function Ψ, this is the time-dependent Schrödinger Eqn. 7.60, with V itself an operator in $V(q, t)$:

$$\left[-\frac{h^2}{8\pi^2 m}\left(\frac{\partial^2}{\partial q^2}\right) + \hat{V}\right]\Psi = \frac{ih}{2\pi}\frac{\partial}{\partial t}\Psi,$$

or

$$\hat{H}\Psi = \frac{ih}{2\pi}\frac{\partial}{\partial t}\Psi = \hat{E}\Psi, \tag{7.69}$$

where $\hat{H}$ is the 'Hamiltonian operator' for the total energy as expressed on the left side. (H is conventionally used in recognition of Hamilton's contributions in mechanics.)

Substituting $\hat{p}^2$ and $\hat{E}$ back into the left side of Eqn. 7.69 gives

$$\left(\frac{\hat{p}^2}{2m} + \hat{V}\right)\Psi = \hat{E}\Psi, \tag{7.70}$$

which is clearly the quantum version of the **Classical Energy Equation** (Eqn. 7.66).

7.4. Relativistic Wave Equation: The Dirac Equation

From the outset, Schrödinger and others naturally considered it essential that the wave equation should incorporate the requirements of 'Special Relativity'. Early attempts to do this (see Kragh, 1981, 1984, 1990) produced what became known as the '**Klein–Gordon Equation**' which, however, gave negative as well as positive energy solutions for the free particle — the difficulty encountered by Einstein when he tried to use a normal wave equation of second-order in time. The approach also failed to account for the fine-structure of the hydrogen spectrum, which Sommerfeld had already successfully done in 1915 by using a relativistic extension of Bohr's theory (§5.3). It could not represent an electron orbiting a proton under Coulomb attraction.

An associated problem concerned 'spin', which had been introduced into Bohr's theory to provide an explanation of line-splitting effects in spectra (§5.3). An attempt to deal with this was made by Jordan and Heisenberg, developing an idea by Pauli, to incorporate spin into the time-dependent Schrödinger equation. Although not really successful, it confirmed the general belief by 1927 that spin was closely associated with relativistic quantum effects.

Dirac provided the breakthrough in his 1926(a)(b) and 1927 papers and especially with his 1928 paper on 'The quantum theory of the electron'. Although the full details are lengthy and complex, we can appreciate the essence of it before turning to the specialist books.

The relativistic value for the energy of a free electron is given by Eqn. 2.13, namely

$$E = \sqrt{c^2 p^2 + m^2 c^4}. \tag{7.71}$$

This gives the Hamiltonian for replacing the Hamiltonian in the time-dependent Schrödinger wave equation (Eqn. 7.69), giving

$$c\sqrt{p^2 + m^2 c^2} = \frac{ih}{2\pi} \frac{\partial \Psi}{\partial t}, \tag{7.72}$$

where, in operator terms, p^2 is replaced (using Eqn. 7.67) by

$$\hat{p}^2 = -\frac{h^2}{4\pi^2} \frac{\partial^2}{\partial q^2}. \tag{7.73}$$

As a relativistic wave equation, this result is noticeably unbalanced. In Special Relativity, time is like a fourth dimension, yet here it is present as a first-order differential whilst the spatial coordinates are present as a second-order square root (which Pauli is said to have described as 'mathematically rather unpleasant'!).

Dirac's inspired breakthrough was to decide not to seek a way of introducing time in the above equation as a second-order differential, but to make the square root into the square root of a perfect square. In terms of our symbolism here, this can be expressed as

$$\left(p^2 + m^2c^2\right)^{1/2} = \left(\alpha_x p_x + \alpha_y p_y + \alpha_z p_z + \beta mc\right), \qquad (7.74)$$

where the coefficients α and β proved to be four-by-four matrices.

With the p's represented by the momentum operator (Eqn. 7.64),

$$\hat{p} = -\frac{ih}{2\pi}\frac{\partial}{\partial q},$$

the wave equation (Eqn. 7.72) then becomes first-order with respect to time and space.

The Hamiltonian for the total energy of the electron is then

$$H = \left(\alpha_x cp_x + \alpha_y cp_y + \alpha_z cp_z + \beta mc^2\right), \qquad (7.75)$$

and, put in operator form into Eqn. 7.69, viz.

$$\hat{H}\Psi = \hat{E}\Psi \qquad (7.76)$$

gives the 'Dirac Equation'.

Adding a term for the interaction of the electron with an electromagnetic field yielded an additional term representing a spin magnetic moment of the electron, with a fixed value of the spin angular momentum, $\frac{1}{2}\left(\frac{h}{2\pi}\right)$, as required by experiment. Hitherto, spin had only been postulated (§5.3).

Another achievement of Dirac's theory concerned the interpretation of the $\pm$ sign implicit in Eqn. 7.71, leading (after some initial debate) to the concept of anti-particles. Their existence was confirmed experimentally when, in cosmic-ray experiments in 1932 using a Wilson Cloud Chamber, Carl Anderson found a positively-charged particle which he called a 'positron' and which was identified as a positive electron by Blackett and Occhialini a year later. Anderson was awarded the Nobel Prize for physics in 1936 for his discovery of the positron in cosmic ray research. (The prize was shared with Victor Hess for the discovery of cosmic radiation.)

Dirac's work uniting Special Relativity with quantum theory has been described by his biographer, Helge Kragh, as the greatest of his numerous contributions to physics. As mentioned in §7.3.8, he shared the 1933 Nobel Prize for physics with Schrödinger for "the discovery of new and productive forms of atomic theory".

In 1932, John von Neumann put quantum theory on a more rigorous basis and his book, translated as *Mathematical Foundations of Quantum Mechanics* in 1955, became a classic and played a major rôle in future thinking on the subject.

Dirac's achievement, which had included the successful prediction of electron spin and the existence of the positron, was a first step towards extending quantum mechanics beyond the region of atomic phenomena. However, attempts to extend the methods of quantum mechanics to deal with particles in the broader sense ran into difficulties. It became evident that a relativistic quantum theory of fields was required, in which **particles in general are seen as the quanta of fields.**

The electromagnetic fields of light carry energy and momentum and can cause electric charges to oscillate, but Einstein's 'discovery' of photons seemed to leave the wave model of light as a mathematical device which determined the average propagation of streams of photons. (This can be seen as analogous to Schrödinger's conception of wave mechanics.) In 'Relativistic Quantum Field Theory', fields are real and particles are the momentary manifestations of interacting fields. This led to a widening out of the subject to deal with a range of problems including the fundamental interactions between elementary particles (the 'weak' and 'strong' interactions) and the whole problem of unifying quantum theory, the Special and General Theories of relativity, and gravity. This on-going story is outside the coverage of this book.

Chapter 8

Matrix Mechanics

8.1. Introduction

Heisenberg, a pupil of Sommerfeld, recalled in the *Bohr Memorial Volume* (North Holland Publishing Company, Amsterdam, 1967) that his development of a new mechanics was influenced by his first meeting with Bohr when the latter gave a series of lectures in Göttingen in 1922. It led him to be convinced that in the construction of an atomic theory, the position, orbit, and motion of an electron should not be considered, as it is in Bohr's theory, because such features are only inferred.

To define the position of an electron in an atom one would have to observe the atom with the light of such a short wavelength (large energy) that the electron would recoil and its position be lost — a notion that was to be embodied in **Heisenberg's Uncertainty Principle** in 1927 (§9.2). Furthermore, in the Bohr theory, the motion of an electron in its orbit strangely plays no part in the phenomenon of the radiation process. [The same type of criticism applied to discussions about de Broglie waves, specifically when they were interpreted in the context of Bohr's orbital atom (Chap. 7).]

In contrast, the frequencies, intensities (and polarisation) of the radiation from atoms, also the energy levels, can be 'observed'. Heisenberg was convinced of the importance of using this approach though he did make the assumption that a spectral line was emitted by what could be regarded as a '**resonator**'. Not surprisingly, the latter exception led to some debate about the meaning of 'observables'!

From discussions at that meeting in 1922, Heisenberg also learned the importance of considering the 'resonator' as an **anharmonic** (i.e. **non-linear**) **oscillator** interacting with a radiation field. Unlike a simple linear harmonic oscillator, it is with an anharmonic oscillator that a change of energy not only alters the amplitude of oscillations but is also associated with a change in frequency (see Appendix H).

These two influences set the scene for Heisenberg's development of a new mechanics which was recognised as a '**matrix mechanics**' by Born and Jordan (1925) and further developed by Born, Heisenberg and Jordan in their extensive '**three-man paper**' (*dreimannerarbeit*) published in 1926, and also by Dirac (1926(a)). It replaced the atomic mechanics of Bohr and in its applications became extremely successful.

Heisenberg's theory, published in the summer of 1925 and often recognised as the origin of the **New Quantum Theory** and **Quantum Mechanics,** was in contrast to Schrödinger's **wave mechanics** published a few months later as a development of de Broglie's ideas about the wave nature of the electron. Schrödinger and others subsequently showed that the two were in fact equivalent. [Wave mechanics is not as pictorially useful as it sounds. In most applications, it would be referring to a hyperspace of many dimensions, and wave amplitudes would determine probability not intensity (as in sound). Both forms of mechanics can be, and are, used for the same types of problem and give the same results, though wave mechanics is easier for energy levels and matrix mechanics for intensities, polarisation etc. Sometimes they are combined.]

At the time when Heisenberg took up the problem, Bohr had introduced the concept of a '**Correspondence Principle**' in which there is a region of scale in which quantum laws match classical laws — matching classical orbital frequencies with spectral emission frequencies at large quantum numbers (§5.3). Bohr invoked this idea again in an attempt, inspired by Einstein's work on transition probabilities (Chap. 6), to determine the intensities of spectra by seeking a relationship between (a) the classical frequencies of the harmonics of a **Fourier series** representation of an anharmonic oscillator and (b) the corresponding frequencies emitted according to quantum theory. Though unsuccessful in achieving his goal, Bohr did show that the frequency of the light emitted in the $n \to n - \alpha$ transition is, in the classical limit, i.e. for large n, the αth harmonic of the fundamental frequency in the nth quantum state (cf. Chap. 6).

The challenge facing Heisenberg was to obtain a theory involving only observable quantities and which would give answers correct also for small n values, something of a quantum theoretical nature that would correspond to the Fourier component of electron motion in classical theory — with the absolute square of amplitudes playing a fundamental rôle in obtaining transition probabilities and theoretical intensities. The signs were that it should be sought in terms of **Fourier analysis**.

Initially, he tried to arrive at the intensity formula for the hydrogen spectrum by using a Fourier series expansion of elliptic orbits but this proved too difficult and he turned to modelling, as Bohr had, the electron's periodic motion as a classical anharmonic oscillator.

8.2. Heisenberg's Approach

Heisenberg considered the case of a 'resonator' in the form of an anharmonic oscillator, with the constraint, mentioned above, that only observable features should otherwise be incorporated. The familiar anharmonic oscillator equation such as (see Appendix H):

$$\ddot{x} + \omega_0^2 x + \mu x^2 = 0, \tag{8.1}$$

with energy

$$E = \frac{1}{2} m \dot{x}^2 + \frac{1}{2} m \omega_0^2 x^2 + \frac{1}{4} m \mu x^4, \tag{8.2}$$

was to be used, but Heisenberg believed that a new kind of **kinematics** (which deals with motions of bodies but not with the forces causing them) was required rather than a new kind of **dynamics** (dealing with the action of forces on bodies).

The actual meaning of $\dot{x}, \ddot{x}$ etc. in the above equation in terms of position and time, as in electron orbital paths, had to be discarded and their interpretation left open. Nevertheless, the motion of such an electron would be regarded as equivalent to a superposition of harmonic vibrations with particular frequencies and amplitudes and it would emit radiations of those frequencies. In that way the behaviour could be expressed as a Fourier synthesis regardless of the mechanical nature of the oscillation.

In Appendix H, we have described how the Fourier series for various energies of oscillation $E^{(1)}, E^{(2)}, \ldots$ etc. can be set out in an array as follows:

$$
\begin{array}{cccc}
q_0^{(1)} \rule[0.5ex]{2cm}{0.4pt} & q_{-1}^{(1)} e^{-2\pi i \nu^{(1)} t} \rule[0.5ex]{1cm}{0.4pt} & q_{-2}^{(1)} e^{-2\pi i (2\nu^{(1)}) t} & \cdots \\[2ex]
q_1^{(1)} e^{2\pi i \nu^{(1)} t} & q_0^{(2)} \rule[0.5ex]{1cm}{0.4pt} & q_{-1}^{(2)} \text{ etc.} & \cdots \\[2ex]
q_2^{(1)} e^{2\pi i (2\nu^{(1)}) t} & q_1^{(2)} \text{ etc.} & q_0^{(3)} \longrightarrow \\[2ex]
\vdots & \vdots &
\end{array}
\tag{8.3}
$$

The constant term $q_0^{(n)}$ of the Fourier series for oscillator energy $E^{(n)}$ is placed on the diagonal shown and the pairs of harmonics are listed down from the $q_0^{(n)}$ and to the right of it. Thus, the array for the Fourier series for the general energy level $E^{(n)}$ is

$$
\begin{array}{cccc}
q_0^{(n)} & q_{-1}^{(n)} e^{-2\pi i \nu^{(n)} t} & \cdots & q_{-s}^{(n)} e^{-2\pi i (s\nu^{(n)}) t} \quad \cdots \\[2ex]
q_1^{(n)} e^{2\pi i \nu^{(n)} t} \\[2ex]
\vdots \\[2ex]
q_s^{(n)} e^{2\pi i (s\nu^{(n)}) t} \\[2ex]
\vdots
\end{array}
\tag{8.4}
$$

Again, energy would be proportional to $\sqrt{qq^*}$.

Heisenberg sought a relationship between these Fourier components of the anharmonic oscillator and the more complicated quantum description of atomic spectra, with emphasis on retaining the general Fourier form but replacing classical magnitudes by quantum ones.

He invoked the **Correspondence Principle** to propose that the sth harmonic of the nth classical energy state corresponds to the quantum transition $n \to n \pm s$ when n is large. The task would then be to extend the treatment beyond the region of applicability of the Correspondence Principle, i.e. to small n (§8.4).

8.2.1. *Heisenberg invokes the Correspondence Principle*

When the energy of a system drops from $E^{(n+s)}$ to $E^{(n)}$ and n is large and s small, the classical and quantum frequencies are the same. This is expressed using the following notation:

$$s\nu^{(n)} = \nu(n+s, n) \tag{8.5}$$

and

$$q_s^{(n)} = q(n+s, n). \tag{8.6}$$

Since $q_s^{(n)}$ and $q_{-s}^{(n)}$ are complex conjugates (Appendix H, Eqn. H.15) we have

$$q_{-s}^{(n)} = q_s^{*(n)} = q^*(n+s, n). \tag{8.7}$$

For a given n, we noted that q_0 is constant, so we write this as

$$q_0^{(n)} = q(n, n). \tag{8.8}$$

Inserting the time factor into Eqns. 8.6 and 8.7 for $q_s^{(n)}$ and $q_{-s}^{(n)}$, we have

$$q_s^{(n)}e^{2\pi i(s\nu^{(n)})t} = q(n+s, n)e^{2\pi i\nu(n+s,n)t}, \tag{8.9}$$

and

$$q_{-s}^{(n)}e^{-2\pi i(s\nu^{(n)})t} = q^*(n+s, n)e^{-2\pi i\nu(n+s,n)t}. \tag{8.10}$$

To avoid the complex conjugate in the above, we note that (a) a reverse transition $E^n \to E^{n+s}$ would have radiation frequency written as $\nu(n, n+s)$, (b) the two frequencies $\nu(n+s, n)$ and $\nu(n, n+s)$ are equal but opposite since the latter refers to absorption rather than emission (i.e. $\nu(n, n+s) = -\nu(n+s, n)$), and therefore (c) Eqn. 8.10 above can be written as

$$q_{-s}^{(n)}e^{-2\pi i(sv^{(n)})t} = q^*(n+s,n)e^{2\pi i v(n,n+s)t} \tag{8.11}$$

$$= q(n,n+s)e^{2\pi i v(n,n+s)t}, \tag{8.12}$$

with

$$q^*(n+s,n) = q(n,n+s), \tag{8.13}$$

and

$$q(n+s,n) = q^*(n,n+s).$$

Collecting Eqns. 8.8, 8.9, 8.12 together, we have

$$\left. \begin{array}{l} q_0^{(n)} = q(n,n), \\[2mm] q_s^{(n)}e^{2\pi i(sv^{(n)})t} = q(n+s,n)e^{2\pi i v(n+s,n)t}, \\[2mm] q_{-s}^{(n)}e^{-2\pi i(sv^{(n)})t} = q(n,n+s)e^{2\pi i v(n,n+s)t}. \end{array} \right\} \tag{8.14}$$

Using the quantum notation on the right of Eqn. 8.14, the array 8.4 can be written as follows:

$$\begin{array}{llll} q(1,1) & q(1,2)e^{2\pi i v(1,2)t} & q(1,3)e^{2\pi i v(1,3)t} & \cdots \\[2mm] q(2,1)e^{2\pi i v(2,1)t} & q(2,2) & - & \\[2mm] q(3,1)e^{2\pi i v(3,1)t} & - & q(3,3) & \\[2mm] \vdots & & & \end{array} \tag{8.15}$$

where

$$\left. \begin{array}{l} q(n,k) = q^*(k,n), \\[2mm] v(n,k) = -v(k,n). \end{array} \right\} \tag{8.16}$$

and

This is known as a '**Heisenberg array**'.

Remember that equating the classical conditions on the left of Eqn. 8.14 with quantum conditions on the right is only valid when the Correspondence Principle applies. At this stage, the numerical values of the terms in the array 8.15 are completely unknown and it gives no information about the quantised motion of an electron performing as an oscillator outside the range of the **Correspondence Principle**. We turn to that in the next section.

8.2.2. *Outside the range of the Correspondence Principle*

In the Bohr theory, a transition from an orbital stationary state can in general involve transition components from various stationary states. To deal with this, Heisenberg accepted Ritz's Combination Principle, (Eqn. 5.21). For example, if we consider atomic orbital transitions $n \to n - \alpha$ and $n - \alpha \to n - \alpha - \beta$, then the transition $n \to n - \alpha - \beta$ also occurs and has a frequency given by

$$\nu(n, n - \alpha - \beta) = \nu(n, n - \alpha) + \nu(n - \alpha, n - \alpha - \beta). \qquad (8.17)$$

The frequencies in a Heisenberg array (8.15) therefore had also to obey relations such as this away from the region of the Correspondence Principle.

Because it was no longer possible to sum the terms as a Fourier series to express the behaviour of the electron in terms of q (and similarly for p) for any given energy, it seemed that the coordinates q and p no longer played a recognisable part in the theory. However, the whole array taken as a single entity did constitute Heisenberg's quantum version of the classical q, and similarly for p, which would be used with Hamilton's mechanics.

There is another way in which the situation was more complicated. In the classical case where the coordinate, q say, of an anharmonic oscillator particle may be expressed as a Fourier series, calculating the energy of the particle would involve squaring this Fourier series to obtain q^2 (cf. Eqn. 8.2), but (as can easily be shown) the squared Fourier series would still be in terms of the same frequencies as the original. To meet the limiting case of the Correspondence Principle, Heisenberg's theory had also to give the same results as the classical theory. Therefore, it was necessary that under all conditions, the square of an Heisenberg array should be an array containing exactly the same frequencies.

Heisenberg's identification of a coordinate with an entire array was revolutionary and initially incomprehensible. He had to develop rules of '**symbolic multiplication**' in which the normal methods of $\times$ and $\div$ used in manipulating q's and p's in classical mechanics did not apply: in particular $pq \neq qp$, which puzzled him.

All this was seen in 1925 by Heisenberg's mentor Max Born and by Dirac to mean that these arrays were **matrices** — hence '**matrix mechanics**' — whose manipulation was already known in algebra and where, for example, multiplication is non-commutative (Born, 1926; Dirac, 1926(a)).

However, the significance of the difference between pq and qp was not to be found in **matrix algebra** but it did emerge by invoking Bohr's quantum condition.

In Appendix C, we see that Bohr's '**quantising condition**' can be expressed as (Eqn. C.20),

$$\oint p.dq = nh, \tag{8.18}$$

where $n = 1, 2, 3 \ldots$ As an expression of the physical restriction on the p, q of possible electron orbitals, the task was to incorporate this into the philosophy of the Heisenberg arrays. To do that, we rewrite the above equation as

$$\oint p\dot{q}.dt = nh, \tag{8.19}$$

and then insert into this the Fourier expansions for p and q. It lead to the fundamental relation known as the '**commutation rule/relation**' (really a non-commutation rule!),

$$pq - qp = \frac{h}{2\pi i}I, \tag{8.20}$$

where I is the unit matrix.

Thus we have

$$pq - qp = \begin{vmatrix} \dfrac{h}{2\pi i} & 0 & 0 & \cdots \\ 0 & \dfrac{h}{2\pi i} & 0 & \cdots \\ 0 & 0 & \dfrac{h}{2\pi i} & \cdots \\ \vdots & \vdots & \vdots & \end{vmatrix}. \tag{8.21}$$

As one of the fundamental statements of matrix mechanics, this defines a mathematical restriction that has to be imposed on the solution matrices of q and p for any (in our case, one-dimensional) mechanical system. (Here it is also the **matrix analogue of Bohr's quantum condition.**)

Note the presence of h as well as the physical restrictions involved. Note also that when this theory enters into the region of the **Correspondence Principle,** the relative importance of h decreases in the limit giving

$$pq - qp = 0, \tag{8.22}$$

as expected in Classical Theory.

Heisenberg's matrix mechanics gave values of frequencies and relative intensities of spectral lines which were the same as those obtained by Schrödinger's wave mechanics.

The use of the anharmonic oscillator by Heisenberg as the hypothetical model for the development of his mechanics tends to attract few comments. The brilliance of choosing it was demonstrated by the success of the resulting theory to which it led. The same had applied in a very similar way when Planck chose the harmonic oscillator for his 'resonators' to represent the sources of black-body radiation regardless of how such radiation was transmitted (§3.2.1).

Heisenberg was awarded the 1932 Nobel Prize for physics for his establishment of quantum mechanics.

8.3. The Reconciliation of Matrix Mechanics and Wave Mechanics

The reconciliation of matrix mechanics and wave mechanics was immediately shown by Pauli (though unpublished), and by Schrödinger in his initial papers of 1926. Schrödinger showed that the solutions of his differential equations, the 'eigenfunctions', were the components of Heisenberg's matrices. (Eigenfunctions are solutions of differential equations and each corresponds, here, to a different state of a system: the 'eigenvalues' then correspond to the energy levels within a given state.) The equivalence of the two mechanics was also soon demonstrated by others, notably by Dirac.

We can have a glimpse of the intimate relationship between the two theories in the following way.

Multiply the Schrödinger wave function Ψ by any function, z say, of q (or p or both) and differentiate the result with respect to (in this example) q,

$$\frac{\partial}{\partial q}(z\Psi) = \left(\frac{\partial z}{\partial q}\right)\Psi + z\frac{\partial \Psi}{\partial q}. \qquad (8.23)$$

Rearranging this gives

$$\frac{\partial}{\partial q}(z\Psi) - z\frac{\partial}{\partial q}\Psi = \left(\frac{\partial z}{\partial q}\right)\Psi, \qquad (8.24)$$

and therefore

$$\frac{\partial}{\partial q}z - z\frac{\partial}{\partial q} \equiv \left(\frac{\partial z}{\partial q}\right), \qquad (8.25)$$

where $\equiv$ means 'is equivalent to'.

Now in wave mechanics, we saw in Chap. 7 that the operator representing momentum was given by Eqn. 7.63,

$$p = -\frac{ih}{2\pi}\frac{\partial}{\partial q}.$$

Equation (8.25) can therefore be written as

$$\frac{2\pi i}{h}(pz - zp) \equiv \left(\frac{\partial z}{\partial q}\right),$$

and when $z = q$, we therefore have

$$\frac{2\pi i}{h}(pq - qp) \equiv I,$$

which is the fundamental statement of matrix mechanics that we first had in Eqn. 8.20.

8.4. Dirac and Matrix Mechanics

In November 1925, Dirac submitted his famous paper 'The Fundamental Equations of Quantum Mechanics' (Dirac, 1926(a)) in which he showed, among other things, the matrix basis of Heisenberg's mechanics. Dirac's biographer, Helge Kragh, recounts how Dirac, on being shown the page-proofs of Heisenberg's paper, said that he had found the treatment not only complicated and unclear but also did not take relativity into account. He felt that it should be possible to state the theory in a Hamiltonian scheme that would conform with the theory of relativity. After several abortive attempts, he made important progress when he saw a resemblance between Heisenberg's non-commutating variables and non-commutating **Poisson brackets** which can be used to formulate Hamiltonian dynamics. Dirac said that after learning more about Poisson brackets, "he could now proceed to formulate the fundamental laws of quantum mechanics in its Poisson bracket formulation". He showed that this new scheme of quantum algebra could be used to specify stationary states and the 'frequency condition' $A_2 - A_1 = h\nu$ (Eqn. 5.9) of Bohr's **Old Quantum Theory** where they had only hitherto been postulates.

Dirac set about seeking to generalise Born's probabilistic interpretation of Schrödinger's wave mechanics and to relate it to his understanding of matrix mechanics. The result was his '**Transformation Theory**', completed in December 1926 and published in 1927. Thoughts along the same lines had also been occupying Fritz London, and also Pauli (in collaboration with Heisenberg), but with less positive outcomes. The rudiments of transformation theory are, however, recognisable in the Born–Heisenberg–Jordan paper (the '**three-man paper**' referred to in §8.1) of 1926 (received for publication in November 1925).

In contrast to Dirac's work, which was based on relating the solutions (eigenvalues) of Schrödinger's wave equation with Heisenberg's matrices, Jordan simultaneously published a transformation theory (1927) that had the same generality and scope and contained essentially the same results as Dirac's, but was based on a strongly probabilistic approach.

What became known as the '**Dirac–Jordan Transformation Theory**' provided an important unification of wave mechanics and matrix mechanics. The details are very complex but the above gives an indication of the context of its origin. It was further developed by these workers and, notably, by John von Neumann (see Bibliography).

Chapter 9

Complementarity, the Uncertainty Principle, and the Copenhagen Interpretation

9.1. Introduction

The 'Copenhagen Interpretation' is so named because it grew from discussions that started in Copenhagen in September 1926 in Bohr's Institute for Theoretical Physics in the University in Copenhagen. Heisenberg was working there at the time as Bohr's assistant, and Schrödinger was there briefly as a visitor. The discussions were an attempt to reconcile the conflicting quantum theories that we have seen had appeared by then. No resolution was achieved but during the following twelve months there were two important, inter-related, developments. One was Bohr's idea of '**complementarity**', used initially to deal with the problem of the wave–particle duality, and the other was Heisenberg's **Uncertainty Principle**, which he found to be a consequence of his matrix mechanics. A lecture by Bohr in Como in the September of 1927 surveyed and attempted to reconcile all this, and one can say that although rather inexplicit, the situation left at that time was regarded subsequently as the 'Copenhagen Interpretation'. We will discuss more of this later.

In the Copenhagen discussions, Heisenberg and Bohr were both committed to quantum jumps and discontinuities but there were serious differences between their views. Unlike Heisenberg, whose matrix mechanics was based mainly on involving 'observables', Bohr was more interested in the relationship between theory and experiment and believed that the concept of wave–particle duality had to form the basis of any discussion of radiation and matter. This was a recent conviction for Bohr. In §5.3, we saw how he had staunchly opposed Einstein's concept of light quanta, how in the 'BKS paper' of 1924, he had attempted to defend his view, and how the argument in that paper was disproved by the convincing evidence of light quanta as demonstrated in the Compton Effect. In a postscript to a paper published in 1925, he referred to this change of view. He developed it further in a lecture to the Copenhagen Academy in December 1926 (and referred to in the journal *Nature* in 1927). He had become very intrigued by the way in which the equation for Einstein's light quanta and de Broglie's equation for matter particles,

$$E = hv, \quad p = \frac{h}{\lambda}, \tag{9.1}$$

each combine particle and wave features, with energy/momentum on one side and frequency/wavelength on the other. He realised that in these equations, Planck's constant, h, establishes a relation between two descriptions that are *mutually exclusive yet jointly essential*. He concluded that a new approach for understanding this was needed: the wave–particle pictures should be seen not as contradictory but as complementary. [We recall that in 1909 Einstein had already foreseen the need for a 'kind of fusion' of wave and particle in his work on radiation (§4.2).]

This was the origin of what became Bohr's Principle of Complementarity which he subsequently called the 'Quantum Postulate', which meant that *if we can prove corpuscular character in an experiment, then it is impossible, at the same time, to prove a wave character: and conversely.* Bohr had partly been inspired to formulate this pragmatic concept by the writings of the American philosopher William James. He felt that the situation being dealt with by his complementarity was analogous to the situation concerning the difficulties in the formation of human ideas inherent in the distinction between subject and object, an important issue dealt with by James in his monumental book *Principles of Psychology* (1890).

Complementarity was to take on a wider meaning in the context of Heisenberg's **Uncertainty Principle**, to which we turn in the next section.

In contrast to Bohr's and Heisenberg's theories, Schrödinger was convinced about continuity in Nature as expressed in his wave mechanical theory, a theory in which 'particles' played no part (they 'existed' only as wavegroups). Schrödinger, who was only visiting Bohr's department, departed having failed to convince Bohr and Heisenberg about his wave mechanics. As Heisenberg recalled later (Heisenberg, 1971), Schrödinger said (in translation) that "If all this damned quantum jumping were to stay, I would regret I ever got involved with quantum theory" to which Bohr replied "But the rest of us are extremely glad you did because you have contributed so much to the clarification of the quantum theory".

Bohr and Heisenberg remained adamant about discontinuities and quantum jumps, and were left to set about trying to find a rational, physical, interpretation of the conflicting views of the various theories. Long discussions and arguments produced no clear resolution.[g] In early 1927, Bohr went skiing for two weeks in Norway and Heisenberg stayed in Copenhagen, a temporary separation that enabled them separately to pursue their thoughts more fruitfully perhaps than might have otherwise been the case. (The extent of the wide-ranging discussions and arguments around this time between the main characters and others can only be appreciated fully by reading their scientific biographies as well as all the papers.)

It was while Bohr was away that Heisenberg drafted his **Uncertainty Principle**, or **Indeterminacy Principle** as it is also known, which in essence means that one cannot, at the same time, measure with unlimited accuracy e.g. the position and the momentum of a particle (§9.2).

---

[g] Ironically, on another occasion, in 1941 during the Second World War, Heisenberg travelled from Germany to have quite different discussions with Bohr in Copenhagen, this time, it seems, about the development of atomic weapons — though it is not known from what point of view (see Cassidy, 1992). The episode was made into a speculative drama for the theatre, based on the book by Michael Frayn in 1998.

When Bohr arrived back in Copenhagen, there were further discussions that led Heisenberg to include and debate, in the paper he was drafting, Bohr's criticisms that were based on the latter's wave–particle duality views. Heisenberg had also sent a draft to Pauli who helped to heal the rift with Bohr and who played an important role in the developing ideas. In March 1927, Heisenberg finally submitted a very comprehensive (27-page) paper for publication. Then, in September of the same year, Bohr gave a wide ranging lecture at the prestigious Volta Centenary Meeting in Como, a meeting held to commemorate the 100th anniversary of the death of Volta who was born and died in Como. The lecture was entitled 'The quantum postulate and the recent development of atomic theory' and, in addition to surveying the various developments, he presented, in rather diffuse terms, his concept of complementarity. Leading physicists from around the world attended, but not Einstein (although he is said to have been invited), or Schrödinger, or de Broglie. (The lecture was published in the journal *Nature* the following year.)

It was the overall situation as seen after the events of those twelve months that became known as the '**Copenhagen Interpretation**' (§9.3). However, it was not defined explicitly at that time, or subsequently, and it was soon to be challenged (§9.4). For the moment, we need to note some details of Heisenberg's Uncertainty Principle paper.

9.2. Heisenberg's Uncertainty Principle

In Heisenberg's very extensive paper referred to above, which launched his Uncertainty Principle, and in his subsequent writings, Heisenberg drew a comparison between the situation in quantum mechanics with what had already happened in the development of relativistic mechanics. Classical mechanics and thermodynamics had used the concepts of force, mass, space, and time, to deal with continuous processes and causality, and they did so by retaining an objective reality that is essentially independent of an observer. To deal with very high speeds and cosmological distances, Newtonian mechanics was replaced by relativistic mechanics. Similarly, another mechanics was now needed to deal with the discontinuities and quantum jumps that had become evident at the very small scale of the workings of atoms.

Since the latter are not directly recognisable at the macro, everyday level, it was not possible to give pictorial descriptions using the usual concepts of position, velocity, etc. A new interpretation of the mathematics of the new mechanics was needed to link the behaviour on the atomic scale to the laboratory-scale observations of atomic phenomena.

To pursue this line of thinking, Heisenberg noted that whereas in the everyday world where the position, q, and momentum, p, of a mass particle follow the basic equations of classical physics, in the quantum world of discontinuities the relation between mass and velocity is given by his **Commutation Relation** (Eqn. 8.20),

$$pq - qp = -\frac{ih}{2\pi}. \tag{9.2}$$

This clearly showed that the physical interpretation of the concepts of momentum and position had to be changed. What, for example, was now to be understood by the 'position' of an object? To answer this, Heisenberg pointed out that viewing an object requires illumination, and the accuracy of the measurement is determined by the wavelength of the illumination: the shorter the wavelength the more precise would be the determination of position. With something as small as e.g. an electron, one would need to use very short wavelength illumination, so Heisenberg considered a hypothetical '**γ-ray microscope**' — a '**thought (*gedanken*) experiment**'. At the instant of observing the electron's position by the scattering of a γ-ray photon, the momentum of the electron would suffer a discontinuous change (the **Compton Effect**). This change in momentum would be greater the smaller the wavelength (the greater the energy) of the light used. Hence, the more precisely the position is determined, the less precisely is the electron's momentum known, and vice versa.

Let Δp be the precision with which the momentum of the electron is determinable due to the change of momentum in the Compton scattering by a γ-ray photon, i.e. $\Delta p \sim h/\lambda$. Also, let Δq be the precision with which the position q of the electron is known at that instant i.e. $\Delta q \sim \lambda$. Combining these gives

$$\Delta p . \Delta q \sim h, \tag{9.3}$$

which is a first expression of the **Uncertainty Principle** (Heisenberg, 1927).

Heisenberg was making the very important statement to the effect that, for any classical theory description of a mechanical system, although there is an exact analogue at the quantum level in atomic processes, there is an indeterminacy in simultaneously measuring conjugate (complementary) variables such as p and q as above, or energy and time as follows:

$$\Delta E.\Delta t \sim h. \tag{9.4}$$

Bohr, anxious to retain his recent conviction about wave–particle duality in any interpretation of quantum formulation (§9.1), argued that Heisenberg's treatment of his hypothetical γ-ray microscope was incorrect. Surely the uncertainty arose from the limited resolving power of the microscope due to its finite aperture in accepting the waves of the light scattered by the electron. He showed that this led to the same expression.

In his 1927 paper, Heisenberg accepted Bohr's criticism and the alternative way of arriving at the equation, and he discussed it and explored it further, producing other derivations, in his subsequent writings (see his 1930 book). He acknowledged that the uncertainty arises from attempts to encompass, simultaneously, phenomena that arise from both wave and corpuscular aspects. In turn, Heisenberg's work prompted Bohr to develop further his ideas about complementarity, which he described in his December 1926 lecture, referred to in §9.1.

It is necessary to emphasise that the Uncertainty Principle is not merely a result of practical limitations in making measurements of, say, p and q. It is only when one property is observed/measured that that property becomes definite, and similarly for the other: unobserved, the particle has no definite p or q.

A simple way (Bohr 1927, 1928) of approaching the Uncertainty Principle, on the wave–particle basis, is to consider a wavepacket representing a mass particle. Let the wave-packet have length Δq. This then represents the uncertainty in determining the exact position of the associated particle. Its length is proportional to the associated wavelength range, $\Delta\lambda$ say, comprising the wave-packet, and from the **Bandwidth Theorem**, we have

$$\Delta q \approx \Delta\lambda. \tag{9.5}$$

Now $\Delta\lambda$ also relates to the uncertainty in evaluating the momentum of the particle, since de Broglie's Eqn. 7.16 tells us that

$$p = \frac{h}{\lambda}.$$

Therefore,

$$\Delta p = \frac{h}{\Delta\lambda}. \tag{9.6}$$

Thus, like Bohr, we arrive at Eqn. 9.3,

$$\Delta p.\Delta q \approx h.$$

In his paper, Heisenberg went on to consider other things such as the path, velocity, and energy of an electron. For example, he considered the concept of an electron orbit in an atom. By 'electron orbit' he said, "we mean a series of points at which the electron occupies one after another as 'position'". However, illumination of a sufficiently short wavelength to 'measure' the orbit would result in the ejection of the electron from its orbit: any sense of orbit would be lost in the process of observing it. The only way of 'detecting' the whole orbit would be to envisage making repeated measurements on a single atom — or, more realistically, measurements on an ensemble of the atoms. This would amount to a probability distribution function for the position of the electron, and here Heisenberg was acknowledging its relationship to Born's probability interpretation of the Schrödinger wave function — adding to Heisenberg's alienation from Bohr. At each 'measurement' there is what became known as a **'collapse'** of the wave function Ψ (cf. §7.3.9) and this is also implicit in Heisenberg's paper.

Heisenberg also sought whatever information could be obtained from transformation theory (§8.4) about the relation between the statistical values of p and q. He knew from the work of Dirac, Jordan, and Born that there is a characteristic statistical connection between quantum theory and classical theory. He obtained the same result, and the Uncertainty Principle can therefore be seen also to have its origins in the Dirac–Jordan **transformation theory**.

More precise derivations of Eqn. 9.3 gave

$$\Delta p.\Delta q \geq \frac{h}{2\pi}, \tag{9.7}$$

and similarly for Eqn. 9.4, see e.g. Heisenberg (1930).

Dirac used Planck's constant divided by 2π as his expression for 'h', and to make the distinction, his 'h' was subsequently written as $\hbar$. The above is then

$$\Delta p.\Delta q \geq \hbar. \tag{9.8}$$

9.3. The Como Meeting and the 'Copenhagen Interpretation'

Although the 'Copenhagen Interpretation' was the name later given to the general outcome of the discussions started in Copenhagen in September 1926 and culminating in Heisenberg's Uncertainty Principle and Bohr's lecture in Como in September 1927, it was not defined explicitly by anyone at that time, or since, due to a lack of a consensus. As a result, there have been extensive discussions in the literature over the years about its meaning (see Stapp, 1972; Fine, 1986, 1996; Jammer, 1989; Cassidy, 1992; Baggott, 2004).

Apart from the lack of a clear definition, even the broad basis of the Copenhagen Interpretation was strongly opposed and challenged, notably by Einstein. We return to that as a separate issue in §9.4.

The problem of definition was partly due to the fact that the writings of Bohr and Heisenberg themselves did not give an unambiguous picture of the logical structure of their position, though it was their work that indisputably formed the basis of the Copenhagen Interpretation. Bohr's writings were generally rambling and verbose, and his crucial lecture at the Como meeting was difficult to comprehend fully.

In §9.2, we have seen that in his paper on the Uncertainty Principle, submitted in March 1927, Heisenberg had attempted to accommodate Bohr's wave–particle concept. He accepted that 'uncertainty' could be seen as arising not exclusively from particle or wave properties, but from expecting both features to be manifest simultaneously.

Bohr discussed this as part of his Como lecture. He saw Heisenberg's **Uncertainty Relations** (Eqn. 9.7),

$$\Delta p . \Delta q \geq \frac{h}{2\pi},$$

and (cf. Eqn. 9.4)

$$\Delta E . \Delta t \geq \frac{h}{2\pi},$$

(9.9)

as each combining particle and wave features (momentum p and energy E with particles, and space q and time t with waves), in accord with his **Complementarity Principle**. The equations give the limits to the simultaneous observation of complementary variables, and observation/measurement forces a choice between one side or the other of the dualism causing the system to be disturbed. This is discussed further in Chap. 10.

At the Como meeting, Heisenberg agreed with Bohr's views. Although hardly forming an explicit definition, many were to regard this ill-defined accommodation of apparently conflicting views as expressing the 'Copenhagen Interpretation'.

However, an unsatisfactory state of affairs remained, e.g. what happens at the instant of a measurement/observation? Heisenberg had considered this in his Uncertainty Principle paper when he argued (§9.2) that in hypothetically 'observing' an electron orbit, the electron would be ejected from the orbit by the use of illumination of sufficiently short wavelength for locating it. Repeated measurements on an atom or, more realistically on an ensemble, would metaphorically 'catch' the electron at different points in the orbit, but the result would amount to a probability distribution. We noted that Heisenberg acknowledged the connection here with Born's probabilistic interpretation of Schrödinger's wave function — much to Bohr's displeasure. Furthermore, we noted in §7.3.9 that the idea of '**wave-function collapse**' associated with the instant of measurement/observation grew from such considerations, though it was a concept that bedevilled theories for a long time (§9.4).

It is not surprising, therefore, that writers have regarded the 'Copenhagen Interpretation' variously as (i) essentially Bohr's Complementarity Principle, (ii) totally implicit in Heisenberg's extensive Uncertainty Principle paper, (iii) implicit in Born's interpretation of Schrödinger's wave mechanics, and (iv) an amalgam of these. Whichever way one looks at it, the Copenhagen Interpretation of quantum mechanics denies reality to anything except the results of observation/measurement. In the macro-world, measurement has a passive rôle, but in the micro-world of quantum theory, it has a more profound rôle in which the act of observation/measurement determines the behaviour found.

Einstein was not present at the Como meeting and he first heard of the deliberations there when he attended the 5th Solvay Conference in Brussels a few weeks later. It was there that he voiced his view of the inadequacy of the current state of quantum theory. Before turning to that, it will be helpful to have in mind how the Copenhagen Interpretation dealt with the perennial guinea pig in these matters, viz, the double-aperture experiment (§2.1).

In Young's historic experiment (Fig. 2.1) of 1801, the observation that bright bands occurred on the screen only when both apertures were open was interpreted as an interference effect due to light having a wave nature. Then, in 1909, G. I. Taylor showed that with a very weak light source, the bands were seen to consist of a build-up of 'spots of light' indicative of light consisting of the particle-like quanta deduced by Einstein in his 1905 explanation of the photoelectric effect (§4.3).

This was a problem needing resolution. In Young's experiment, the profiles of the bright bands observed on the screen were very satisfactorily explained as being due to interference of wave trains spreading out from the two apertures: they reinforce on the screen where they superimpose in phase and cancel where they superimpose out of phase. It was hardly possible to explain this in terms of light consisting of photon 'particles' as implied by Taylor's experiment with a weak source. A photon from the light source could hardly be expected to have passed through both apertures, arrive at the screen, and according to where it arrived, annihilate itself or not. In 1961, Claus Jönsson in Tübingen demonstrated the double-aperture experiment with electrons for the first time, and in 1974 in Bologna, P. G. Merli *et al.* demonstrated it with single electrons. Since then, particle interference has been demonstrated with neutrons, atoms, and small molecules (§2.1).

So how did the Copenhagen Interpretation deal with this? How does each electron, for example, leave its source as a particle and arrive on the screen as a particle, having passed through both apertures in order to interfere (with itself) and know where to arrive on the screen and contribute to the build up of the interference pattern? Each particle passing through the apparatus would not even know whether or not both apertures are open. Furthermore, why don't the electrons all follow the same path and arrive at the same spot on the screen?

It seemed evident that a particle such as a photon or electron travels through the experimental set-up as a wave but arrives as a particle. The 'wave', however, was regarded at this time as a probability wave describing the probability of finding the particle at any particular place.

Arriving at the screen from the two apertures (Fig. 2.1), there would be constructive or destructive interference at different points on the screen, depending on the path differences involved. Where there is constructive interference, the particle is observable — the probability wave collapses to become a definite particle at the instant of observation. If arrangements are made to actually observe particles going through the apertures, then the process of observation destroys the interference on the screen.

Thus, the particles are aware of the conditions throughout the space of the experimental set-up. Such 'non-locality' led Einstein later to refer to such phenomena as 'spooky action at a distance', a concept that troubled him profoundly (§9.4) though it was proved correct in the 1980s (Chap. 10).

Bohr argued that the whole experimental set-up including the observer has to be considered. When the apparatus is used to observe waves, interference effects are observed, and when used to monitor the passage of particles then particle behaviour results.

Clearly, this accommodates most if not all of the views concerning wave–particle duality, complementarity, and the probabilistic interpretation of wave-mechanics (though Born himself did not go as

far as to consider '**wave-function collapse**'). The Copenhagen Interpretation (including wave function collapse if not in name) is also seen
as implicit in Heisenberg's comprehensive 1927 paper which included
his **Uncertainty Principle**. Bohr recognised the latter as giving the limits (via Eqn. 9.9) to which simultaneous observation of **complementary** variables can be made. Heisenberg had shown that one can make
measurements which observe, accurately, the position of a particle,
or one can accurately measure its momentum, but not both at the
same time. Until a measurement is made, the 'particle' has neither a
definite p or q, a point that was emphasised in §9.2 as not reflecting
imperfection in measurement. The 'quantum object' simply does not
have known precise momentum or position until it is observed — and
then not both. Fortunately, such effects only become evident for very
small entities (photons, electrons, etc.) because of the minute value
of Planck's constant, the controlling factor in the formulation of the
Uncertainty Principle Equation (Eqn. 9.9).

It will be noticed that Bohr had not explicitly defined '**complementarity**', but he was describing how complementary modes of
description were adequately dealt with by the formalism of quantum
theory. The question of explicit definition was the subject of much
discussion between Bohr, Pauli, and others (see Jammer, 1989 — see
Bibliography).

Despite all the philosophical problems it posed, the '**Copenhagen
Interpretation**' held sway until the 1980s. Ironically, the vagueness
of some of its aspects may have contributed to its adaptability and
endurance. However, it was challenged, notably by Einstein, though
unsuccessfully, and we next need to see the part that it played in
leading to later developments.

9.4. Copenhagen Interpretation Challenged

9.4.1. *Background*

Within a few weeks of the Como meeting, the essence of the
Copenhagen Interpretation was challenged by Einstein at the 5th
Solvay Conference on 'Electrons and Photons' held in Brussels in

October 1927. The meeting was attended by nearly all the main origi-
nators of the quantum theory, including Einstein this time. It covered
a wide range of topics in physics and in the general discussion towards
the end, Bohr was asked to address the meeting. It was in this way
that Einstein heard for the first time about what had happened at the
Como meeting. Already opposed to the idea of discontinuities and
the statistical nature of the new theories with its lack of **causality** cf.
§1.6, he embarked on proposing, at available opportunities during
the rest of the meeting, various '**thought experiments**' (as Heisenberg had done with his 'γ-ray microscope' (§9.2)) intended to show
the fallacies in the new theories. In each example Bohr proved him
wrong.

The fundamental disagreement between Bohr and Einstein went
back to 1920 with Bohr's reluctance to accept Einstein's concept of
light quanta. It continued for many years and was never resolved.
Jammer (1974) has described it as one of the great debates in the
history of physics, comparable with the Newton–Leibniz controversy
of the early 18th century.

The challenges were mainly undocumented, but Bohr gave a retro-
spective survey in a 41-page article entitled 'Discussions with Einstein
on Epistemological Problems in Atomic Physics' which he wrote for
a book published in 1949, edited by Schillp (1949), in celebration
of Einstein's 70th birthday. In the article, Bohr first set the scene by
surveying the various problems that had accumulated as a result of
Planck's discovery of the quantum of '**action**'. It is useful for us to do
the same here, putting the emphasis as Bohr did, on the philosophical aspects. (A few of the theoretical points he mentioned have not
been dealt with in detail in this book, to avoid going off at a tangent
from the main storyline, but they are mentioned here because they
supported the development of ideas.)

Bohr started by noting that Classical Physics had become an ide-
alisation that only applied in the limit when 'all actions involved'
are large compared with the quantum. In the region where quantum
effects are manifest, causality seemed to have been lost. Was this real
or was it a failure of the new theories?

Planck's discovery of the quantum in his study of black-body radiation in 1900 had been 'guided' by the relationship between the laws of thermodynamics and the statistics of mechanical systems (Chap. 3). Einstein had then shown in 1905 how the photoelectric effect depends on *individual* quantum effects (§4.3). Such a process involves the emission or absorption of individual quanta (photons) with energy and momentum,

$$E = h\nu, \quad p = \frac{h}{\lambda}. \tag{9.10}$$

This concept of the photon posed the problem of being irreconcilable with optical interference effects — effects that, ironically, need the usual wave model to define the concepts of ν and λ in the above equations to give energy and momentum!

The failures of classical physics were accentuated by its inability to explain the inherent stability of the structure of the atom as deduced by Rutherford in 1911. In Bohr's hands, this stability, and the empirical laws governing the spectra of the elements, showed that changes of orbital energy involved transitions between so-called stationary quantum states (§5.3).

Einstein had made important progress in his 1917 work on the statistics of transitions (Chap. 6). It included an independent derivation of Planck's radiation formula and also dealt with the question of the intensities of atomic spectra in Bohr's theory. He emphasised the fundamental nature of statistics in these processes and drew support from the analogous situation in the transformations involved in radioactivity.

In his article, Bohr also emphasised that the involvement of statistical aspects in much of quantum theory had become quite different from the use of statistics in dealing with e.g. mechanical macrosystems. In quantum physics, they pointed in many cases to the actual inability of classical physics itself to deal with events at the elementary level. This is an important difference.

Bohr recalled that problems had continued to abound. One concerned the fact that Einstein's explanation of the **Compton Effect** (§4.2) implied that the direction in which photon momentum is transferred to an atom dictates, unrealistically as he was only too aware, that the resulting emission will be in the opposite direction, i.e. the radiation process is 'unidirected'. Clearly, no simple picture of corpuscular collision was adequate.

Then, in 1923, de Broglie suggested that Einstein's wave–particle duality of light (§4.2) also applied to material particles, and this was confirmed experimentally. [Einstein had already envisaged and used an analogy between properties of thermal radiation and gases (§4.3).] de Broglie's idea was radically developed by Schrödinger who, following Hamilton's analogy between mechanics and optics, showed how stationary states of atomic systems were represented by the proper (eigen) solutions of his wave equation.

However, there remained the apparent contradiction between the wave mechanics theories and the individuality of elementary processes.

In the 'matrix mechanics' that Heisenberg, Born, Jordan, and Dirac progressively developed at the same time, the variables of classical mechanics were replaced by symbols subjected to a non-commutative algebra, and which replaced the whole idea of orbits (Chap. 8).

The two mechanics were shown to be equivalent, and general methods of this **'quantum mechanics'** were developed for an essentially statistical description of atomic processes yet incorporating the feature of individuality.

Bohr went on to state that despite the enormous success of the applications of the theory at this stage, its abstract character "gave rise to a widespread feeling of unease".

It was just prior to the Como meeting that Heisenberg pointed out that the knowledge obtainable about atomic systems involves an 'indeterminacy'. This became his **'Uncertainty Principle'** and he showed that his **Uncertainty Relations** (Eqn. 9.9) were central to the paradoxes involved in attempts to analyse quantum effects in terms of the customary physical pictures (§9.2). After criticism by Bohr, Heisenberg acknowledged (and included it in his 1927 paper) that the uncertainty can be seen to arise from attempts to encompass, simultaneously, phenomena that originate with both wave and particle aspects.

At the Como meeting, Bohr advanced the idea that the Uncertainty Relations were in accord with his concept of '**complementarity**'. Choice of a feature to be 'observed' puts a limit on the accuracy available for obtaining information of the conjugate (complementary) feature. Evidence obtained under different experimental conditions cannot, separately, be understood with a single picture. Such pieces of evidence must be regarded as 'complementary', and only their totality gives all the 'possible information' about an object.

This was the background against which Einstein, at the 5th Solvay Conference in October 1927 and again at the 6th Conference in 1930, progressively raised various objections to the general outcome of the Como meeting.

9.4.2. *Einstein's objections*

Bohr had regarded his principle of complementarity as a 'rationalisation of the very ideal of causality' and had hoped that this 'positive element', which he had introduced at the Como meeting in September 1927 and had continued to develop, would satisfy Einstein's strong belief in continuity and causality. However, this was not to be so and, at the 5th Solvay Conference a few weeks later, when Einstein heard for the first time the details of the developments, he voiced his disapproval. He did not make a formal contribution at the meeting (though he had been invited to do so by the chairman, H. A. Lorentz) but he did make a few remarks on the final day about the question of quantum statistics versus complete individual interpretations.

To illustrate his criticism, Einstein took the example of particles (photons or electrons) passing through a small aperture in a screen and giving the familiar diffraction pattern on a second screen (or photographic plate) placed some distance away. He considered it from a statistical standpoint rather than in terms of Heisenberg's Uncertainty Relations, of which he was aware by then. If the wave function represents the probability of a particle arriving at some point on the screen, then at the instant of arrival at that point it cannot at the *same time* arrive at another point (cf. '**collapse of the wave function**'). Moreover, this interpretation violates the **Special Theory of Relativity** because 'signals' cannot travel faster than the speed of light. Alternatively, if the wave function represents the whole cloud (ensemble) of particles travelling through the system, then $|\Psi|^2$ represents the statistical probability of finding one particle of the cloud arriving at a given point. This second scenario seemed more feasible to Einstein and it reinforced his view concerning the statistical nature of the existing quantum theory and its inadequacy in dealing with individual particles.

There is no record of Bohr's response at the time but in his 1949 article he reasoned that Einstein's analysis of the first scenario above was incorrect because the assumption had been made that the aperture and screens were assumed to have exact positions in space and time. There would be no uncertainty in the position and time when a particle leaves the aperture, and the energy and momentum of the scattered particle would, in accordance with the Uncertainty Relations (Eqn. 9.9), be totally indeterminate — giving a spherical wavefront. However, in accordance with those relations, there would be some uncertainty in position and time at the aperture, and he showed that when a particle leaves the aperture, the uncertainty in its energy and momentum would consequently be reduced, and the prediction of where the diffracted particle would hit the screen would be correspondingly increased. Bohr claimed that, within the limits imposed by the Uncertainty Relations, quantum mechanics gave as good a description of individual particle events as one could expect on the atomic scale.

In his reminiscences, Bohr recalled other 'thought experiments' and discussions along the same lines, but such arguments failed to convince Einstein who persisted in believing that it should be possible to construct a better theory of 'individual elementary processes'.

It should be noted here that although at the time of that 5th Solvay Conference, Einstein had, just a few months earlier, in April 1929, received via Bohr a pre-publication copy of Heisenberg's Uncertainty Principle paper (and Heisenberg had been in correspondence with him in the May and June about the Uncertainty Relations), nevertheless at the time of the Conference Einstein's disagreements were not related to the Uncertainty Relations but were much broader in scope (Fine, 1986, 1996). However, by the time of the 6th Solvay Conference in

Brussels in October 1930, Einstein had turned his attention specifically to the Uncertainty Relations. Before looking at the new direction of Einstein's criticisms, we need to note a development in Bohr's thinking that was relevant.

During 1928–1929, Bohr wrote in greater detail about his interpretation of quantum mechanics, especially with regard to his approach to Einstein's Special Theory of Relativity (see Jammer, 1974). Bohr was making the point that Planck's discovery of the quantum of action had presented a situation similar to the discovery of the finite speed of light. To appreciate this, we first note that in the everyday macro-world, the smallness of Planck's quantum of action makes possible a simultaneous space–time causal description of events, and that ordinary speeds make possible the separation of space and time. In contrast, in microphysical processes, the reciprocity/complementarity involved in making measurements cannot be ignored, just as in high speed phenomena there are questions concerning simultaneity of observation. Jammer (1974) recounts how Bohr expressed the view that Heisenberg's Uncertainty Relations "safeguard the consistency of quantum mechanics", and the limit to the speed of light "safeguards the consistency of relativity theory".

Hence, Einstein set out to show that the relativity theory in fact disproves Bohr's contentions. He used the 6th Solvay Conference in 1930 as an opportunity to attempt to do this and dealt specifically with the energy–time form of the Uncertainty Relations (Eqn. 9.9). Again there were only informal discussions during the conference (where the official topic was the magnetic properties of matter) with no records kept, but the so-called '**photon in a box**' thought experiment that Einstein proposed has become a much written about legendary highlight of the long running **Bohr–Einstein debates** (Bohr, 1949; Jammer, 1974). The experiment amounted to visualising a box containing radiation, the box being fitted with a shutter that could be opened for a very short, and precisely known, time by a clock in the box to allow a single photon to escape from the box. The box could be weighed before and after the escape and from the difference, using Einstein's $E = mc^2$, the precise energy of the photon would be known. Such precise measurements of chosen accuracy concerning the time and energy of the passage of a photon from the aperture therefore contradicted the energy–time Uncertainty Relations.

After an allegedly sleepless night, Bohr felt able to rebut Einstein's challenge using, ironically, the latter's own **General Theory of Relativity**. He proposed that to perform the experiment, the box would need to contain a clock (hence the other name, '**clock in a box**', for the experiment) to time the operation of the shutter, and the box could be weighed before and after the release of the photon by suspending it on a spring balance. According to Einstein's relativity theory, the movement of the box on the spring balance as the weight changes due to the loss of the photon changes the time-keeping of the clock. Also, the latter introduces an uncertainty in the precise timing of the opening of the shutter which in turn affects the weighing of the box. Bohr showed that the greater the accuracy of the measurement of the energy of the photon (via $E = mc^2$) the less accurate is the measurement of the timing of its escape. He showed that these were in accordance with Heisenberg's Uncertainty Relations.

Bohr's response to the 'photon in a box' experiment was rejected by some, with alternative explanations offered (Jammer, 1974). At the time, however, the general view was that Bohr had won the debate. Einstein, on the other hand, though accepting that the experiment seemed to be free of contradictions, continued to feel that it was not reasonable. He understood the Uncertainty Relations as a limitation of simultaneous precise measurement, but he did question them as a limit to simultaneous reality.

Einstein's acceptance both of Bohr's criticism and the Uncertainty Relations was evident in his paper with Tolman and Podolsky in 1931, which dealt with a new variation of the 'photon in a box' thought experiment in which experimental information about a particle might be used to make predictions about another particle. This '**ETP paper**' claimed to show that the principles of quantum mechanics, as the subject stood, extended to involving an uncertainty in even the description of past events, analogous to the uncertainty in predicting future events. Not surprisingly, this led to much debate but it was regarded as not memorable in view of the 'EPR' thought experiment that came four years later, to which we turn below. [There were many discussions around this time generally about the logical status of the Uncertainty Relations (Jammer, 1974).] Still with the 'photon in a box' thought experiment in mind, Einstein was now setting about trying to show the *incompleteness* of the theory of which the Uncertainty Relations were a part, rather than trying to disprove them. Ironically, Einstein's own contributions had put emphasis on the relevance and development of statistical methods but, as we have noted earlier in several places, although he believed that they were a useful mathematical device for dealing with phenomena involving large numbers of elementary processes, he realised that they did not give an account of individual processes. Quantum theory was perhaps the correct theory of statistical laws but it did not provide an adequate treatment of individual elementary processes. Furthermore, with an instinctive dislike of the idea of a probabilistic universe in which the behaviour of individual atoms depends on chance, he felt there should be a deeper, independent, theoretical framework — what he called **'objective reality'**— for dealing with the latter.

By now, Einstein was almost alone in having this belief and it was to preoccupy him for the rest of his life. As he put it, his main aim was to find a 'complete theory' in which a "sufficient condition for the reality of a physical quantity is the possibility of predicting it with certainty, without disturbing the system". This sentiment was expressed in the four-page essentially non-mathematical paper published in 1935 by Einstein, Podolsky, and Rosen, which had the title 'Can quantum–mechanical description of physical reality be considered complete?'

In this **'EPR'** paper, Einstein's idea now was to find a way that in principle would obtain knowledge of the state of a quantum particle without disturbing it, thus overcoming the arguments Bohr had previously used to disprove his challenges. The paper considered a **thought experiment** similar to the 'photon in a box' experiment, again involving two quantum particles, but this time where, for reasons we shall see, they have interacted briefly with each other and then moved apart. The essence of the experiment can be summarised as follows.

Consider two particles, A and B say, with momentum and position variables (p_A, q_A) and (p_B, q_B), respectively. They have a definite total momentum $P = p_A + p_B$ which is conserved throughout the experiment, and a definite total position $Q = q_A - q_B$. The particles interact, e.g. by colliding, and separate to an unlimited distance. After separation, suppose the momentum p_A of A is measured. The momentum p_B of B can therefore be deduced without disturbing B in any way: it is an '**element of reality**' pre-existing in the state of B. If, *instead*, q_A had been measured then q_B would become known and that, too, must have been an element of reality of B. Although quantum mechanics does not allow B to simultaneously carry the values of p_B and q_B (cf. '**collapse of the wave function**'), Einstein believed that the particle must carry both values if it is possible to measure either. Also, he did not believe that a remote measuring device could influence the state of a particle.

As Einstein's belief in **local reality** contradicted Bohr's belief in indeterminancy, he felt that there must be **hidden variables** that carry values which determine the *precise* result of a measurement. Furthermore, if the 'reality' of the position or momentum of a particle is only achieved by measurements on another, albeit related particle, then some kind of '**action at distance**' must operate, and this would of course need to be instantaneous because there need be no limit on their separation. That would violate the **Special Theory of Relativity** which restricts communication to being no faster than the speed of light.

The EPR experiment was therefore claiming to show that either quantum mechanics is incomplete as Einstein believed, or there is some sort of instantaneous communication ('**correlation**') over no matter how large a distance. The authors did not believe in such action at a distance because the momentum and position of a particle exist all the time: quantum mechanics was therefore incomplete. As H. T. Flint (1935) immediately commented in the journal *Nature* "The authors seem to prefer the artist's portrayal of the landscape rather than a conventional representation of its details by symbols which bear no relation to its form and colour" (Flint, 1935).

[We may note here that although Einstein and others referred to the experiment as the '**EPR paradox**', this was not strictly a correct description because the experiment demonstrated incompleteness of quantum theory rather than contradiction. An alternative sometimes used is to refer to it as the '**EPR incompleteness argument**'.]

Although the EPR paper included reference to wave function and operator aspects, others looked into the various aspects in more detail (see e.g. Jammer, 1974; Fine, 1996; Baggott, 1992, 2004; Duck and Sudarshan, 2000). However, the general sentiment the authors expressed in their original paper and summarised above has remained, as some authors have put it, its 'lasting message'.

By all accounts, Podolsky wrote the EPR paper without Einstein even seeing a draft, and Einstein regretted some of the wording. The criterion of '**physical reality**' was not essential to the message of the paper, and it was this that was used by Bohr in his seven-page rebuttal, rapidly published afterwards in 1935 under the same title as the EPR paper. However, he failed to counter, successfully, the challenge posed by the EPR paper. He rejected the suggestion that the EPR experiment posed difficulties for the Copenhagen Interpretation, stressing yet again the important rôle of measurement in defining the '**elements of reality**' that can be observed. Measuring the momentum of A with certainty and thereby deducing the momentum of B excludes the possibility of accurately discovering its position by that measurement. The element of '**physical reality**' of particle B to be determined had depended on the type of measurement used on A.

At this point Bohr was obliged, in his argument with Einstein, to drop his previous use of the effect of mechanical/physical interference in making a measurement. He now turned to the vaguer notion of the essential inseparability of observer and observed. One chooses the whole experiment and its measurement and one cannot make a different measurement at the same time. There is no conflict and no valid ground for deducing that quantum mechanics is incomplete.

In contrast to Bohr, Schrödinger congratulated Einstein on the EPR paper and there were several exchanges by correspondence between them. It was then that Einstein expressed his dissatisfaction with some of the sentiments in the paper. In the correspondence, he explained his view of '**incompleteness**'. He used examples to explain that by 'incompleteness', he meant a description where a deduction can only be probabilistic, one that is based on circumstances in which there is a further truth, involving '**hidden variables**', to be found. The states yielded by quantum mechanics would not correspond to exact values of these variables but to specific averages of them. Einstein had used the idea of hidden variables in 1927 in a discussion of Schrödinger's wave mechanics but he did not pursue it in the present context. Although he did not attempt to formulate a 'hidden variable' theory, he supported the concept and remained convinced that a new theory was needed. He spent his remaining years in an (unsuccessful) quest for a **Unified Field Theory** — to be an extension of his General Theory of Relativity to include Quantum Mechanics.

Schrödinger quickly became immersed in the problem. The correspondence continued and in an 'essay' read in October 1935 by Born at a meeting of the Cambridge Philosophical Society (and published in the Proceedings), where he generalised the EPR argument, regarding it as an indication of a serious deficiency in quantum mechanics. At about the same time, he published a three-part paper, with further elaboration, in *Die Naturwissenschaften*. In this, he dealt with the concept of wave functions and their statistics as reflecting an underlying physical reality of waves, wave-packets, and their superpositions.

In the **Copenhagen Interpretation**, the wave function for a two-particle system (A, B) such as we have been considering does not separate as the particles separate. When a measurement is made, there is a '**collapse of the wave function**', though it continues to be spread out over a considerable distance, leading to what Schrödinger called '**entanglement**' of the two bodies — first envisaged by him in 1926 (§7.3.8).

In the EPR interpretation, on the other hand, the two particles are regarded as isolated from each other and are no longer described by a single wave function at the moment when a measurement is made. Unlike the Copenhagen Interpretation, there is now '**local reality**'. The ability of the particles to separate into two, independent, locally real, entities, is often referred to as '**Einstein separability**'.

One of the most interesting points of Schrödinger's three-part paper was his presentation of the famous '**Cat Paradox**'. He wanted to show that uncertainty in variables is ultimately untenable (the title of the fifth section of the paper translates as 'Are the variables really fuzzy?'). Describing it as a *ludicrous example,* he visualised a cat enclosed in a steel chamber that contains a small amount of radioactive material — so small that the probability of it decaying in, say, one hour, was 50%. If it does decay, a Geiger counter in the chamber fires and triggers the release of poison that kills the cat. At the end of the hour is the cat alive or dead? According to the Copenhagen Interpretation, it is neither, until we look inside to see — an interpretation hard to accept, though of course it was extending the quantum mechanics of the micro-world to the macro-world where they do not necessarily apply.

The 'Cat Paradox' had its origins in the EPR experiment and it has been widely discussed over the years. In his 1996 book, Arthur Fine devoted 22 pages to a well-referenced discussion.

However, experimental tests by the team led by Alain Aspect (Aspect *et al.,* 1982) favoured Bohr's Copenhagen Interpretation and the non-existence of hidden variables. It seemed that '**entanglement**' was to be accepted.

A more detailed survey of the above, and the way the theories developed up to the present time, is the subject of the next chapter.

Chapter 10

Indeterminacy and Entanglement

Sara M. McMurry

10.1. Introduction

For nearly three decades after the EPR paper (Einstein, Podolsky and Rosen, 1935) (discussed in §9.4.2) Einstein's disagreement with Bohr concerning the incompleteness of quantum theory was seen as a purely philosophical argument. The **EPR thought experiment** is paradoxical only if viewed from the point of view of Einstein's philosophical position, where a measurable 'element of reality' must be a well-defined property of a particle. On the other hand, according to the **Copenhagen Interpretation**, the properties of a particle are not well-defined — they are indeterminate — until they are measured. In the EPR paper, Einstein highlighted what he saw as two problems with the Copenhagen Interpretation: firstly, the indeterminacy in the value of a measurable variable, and secondly the **non-locality** ('spooky action at a distance') implied by the instantaneous effect of a measurement made on one particle on the state of a second particle, no matter how far apart the particles are at the time.

Then, in 1964, John Bell, an Irish theoretical physicist working at CERN, transformed the argument by deriving a mathematical expression, known as **Bell's Inequality**, susceptible of experimental test, which could discriminate between the approaches of Einstein and Bohr. Any theory satisfying Einstein's requirements of reality and locality must always satisfy this inequality, whereas the predictions of quantum theory violate it in certain situations. The technology for obtaining an unequivocal result from a test of Bell's expression did not exist until the 1980s, when the French physicist Alain Aspect and co-workers performed a series of experiments of the EPR type. The results indicated that Bell's Inequality was violated. More recent experimental tests of Bell's Inequality have all tended to confirm this result.

Aspect's experiments, described in §10.2.4, used entangled pairs of photons and measured their polarisation. In §10.2.1, we show how the quantum idea of indeterminacy applies to the case of photon polarisation, and use this in §10.2.2 to describe the type of EPR experiment performed by Aspect *et al.* In §10.2.3, Bell's work is explained in terms of EPR experiments in which photon polarisation is the measured quantity.

Non-locality is manifest in the **two-slit experiment** (§9.3) since in order for the interference pattern to be seen, the trajectory of each quantum particle must be indeterminate: the particle cannot be pictured as following a unique, though unknown, trajectory through one slit or the other, yet the particle cannot split in two with one half going through each slit. Therefore, a particle, as it passes through the apparatus, is not localised. Any measurement which locates the particle sufficiently to identify which slit it went through will destroy the interference pattern. However, there is a much more serious non-locality in the entangled states of pairs of particles in EPR-type experiments, as Bell's work demonstrates.

Bell's theoretical work has led to a deeper understanding of quantum correlations and entangled states, and to the evolution of a new area of physics — quantum information theory, which we shall discuss in §10.3. Advances in technology mean that quantum information theory may be investigated experimentally, and are leading to practical applications. Current research efforts are focussed on quantum cryptography and quantum computing.

The possibility of entangling the state of a macroscopic object with that of a quantum particle was used by Schrödinger, in his famous **Cat Paradox** (§9.4.2) to explain his dissatisfaction with the probability interpretation of the wave function he had introduced. He was also unhappy with the idea of 'quantum jumps' and the collapse of the wave function on measurement. These difficulties are the focus of quantum measurement theory, which is the subject of §10.4. This branch of quantum theory has made enormous strides during the second half of the 20th century, and is refining the interpretation of quantum theory.

10.2. Entangled Pairs of Particles and Bell's Inequality

In the EPR paper (§9.4.2), Einstein made it clear that he believed that

(i) values of measurable quantities are physically 'real', having existence whether or not they are actually observed;

(ii) the outcome of any measurement cannot be influenced by the settings on a remote detector. (This is his '**locality**' assumption.)

As we outlined in §9.4.2, the original EPR thought experiment concerned the measurement of either momentum or position on one of an entangled pair of particles (particle A), the results of which gave immediate knowledge of the corresponding variable of the other (particle B). Therefore, according to Einstein, if B's momentum is correctly deduced from a measurement on A, this value of momentum must have been carried with B as an 'element of reality', although it was unknown by the experimenter until A's momentum was measured. Similarly if A's position were measured, giving immediate knowledge of B's position, B must have carried information about its position with it. This contradicts the Uncertainty Principle, which implies that a particle cannot carry precise information about both its momentum and its position.

Furthermore, the Copenhagen Interpretation asserts that as soon as the momentum (or position) of A is measured, the momentum (or position) of B also becomes precise, no matter how far apart A and B are at the time. This is what Einstein called 'spooky action at a distance' (§9.3). Einstein's Special Theory of Relativity tells us that information cannot travel faster than c, the speed of light in vacuum, and so it seems that the measurement on A should not be able to bring about an instantaneous change in the state of particle B. According to

Einstein's locality assumption, the measurement on A cannot directly affect the state of B: the result of the measurement on B should depend only on the properties carried by B and the local measuring apparatus. It should be independent of whether the remote apparatus which A enters is set to measure position or momentum. In fact, as we shall see in §10.3, the instantaneous effect on B of the measurement on A does not, by itself, carry information. Hence, in an EPR experiment, the speed of transmission of information remains less than c, but this does not preclude non-local interaction at a distance.

A variation of the EPR thought experiment was proposed by Bohm and Aharonov in a paper in *Physical Review* (Bohm and Aharanov, 1957), in which components of the spin of a pair of particles were measured instead of momentum or position. This version makes the argument very clear, and may also be extended to measurements of the polarisations of a pair of photons, which provides the simplest practical realisation of an EPR experiment.

10.2.1. *Photon polarisation and indeterminacy*

Ordinary light — daylight or light from a conventional source — is unpolarised. However, if a beam of such light shines on a piece of polaroid, only half the light intensity is transmitted. The transmitted light is polarised so that all the electrical vibrations in the light are along the same direction, which is determined by a preferred axis in the polaroid material. A piece of polaroid transmits light polarised parallel to its preferred axis, and blocks light polarised at right angles to that axis. Unpolarised light can be treated as a $50:50$ mixture of light polarised parallel to and perpendicular to any chosen axis.

In the classical theory of electromagnetic waves, this is easily understood. The electrical vibrations can be in any direction in a plane perpendicular to the direction of the beam, and an unpolarised beam contains a random mixture of waves polarised in all possible directions. If a wave polarised in the x direction falls on a piece of polaroid with its axis in the x direction, then it is completely transmitted, while if a wave polarised at right angles to the x direction falls on the

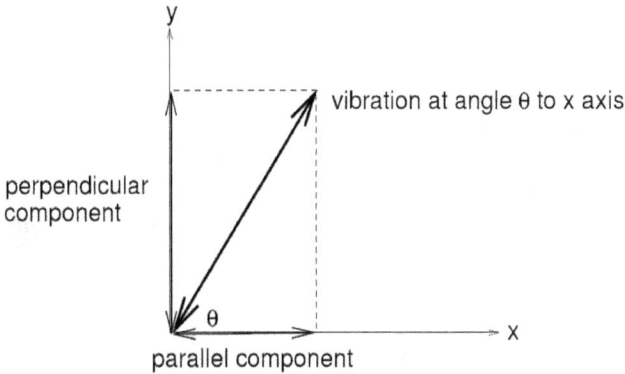

Figure 10.1. A vibration along a direction at angle θ to the x-axis may be resolved into components parallel and perpendicular to that axis.

polaroid, it is completely blocked. For a wave polarised at an angle to the x direction, the x component of the vibration is transmitted and the perpendicular component is blocked. Figure 10.1 shows the components of a vibration at angle θ to the x-axis. The vibrations of all the waves in the beam can be resolved into components parallel or perpendicular to the polaroid axis, and only the components parallel to that axis are transmitted.

When light is treated as a beam of photons, however, the result is no longer so intuitive. If we imagine a photon incident on a piece of polaroid with its axis in the x direction, then we should expect that photon to be transmitted if it is polarised in the x direction and blocked if it is polarised in a perpendicular direction. But what happens if the photon is polarised at an angle to the x-axis? In the case of a wave, one component of the vibration is transmitted and the perpendicular component blocked. But a photon cannot be split in two as it enters the polaroid: either the whole photon is transmitted or the whole photon is blocked. According to quantum theory, the probability of the photon being transmitted is determined by $\cos^2 \theta$, the square of the cosine of the angle θ that its polarisation makes with the x-axis. Similarly, the probability of the photon being blocked is determined by $\sin^2 \theta$. Since $\cos^2 \theta + \sin^2 \theta = 1$, the total probability for one of these two possibilities to occur is one. The photon will definitely be

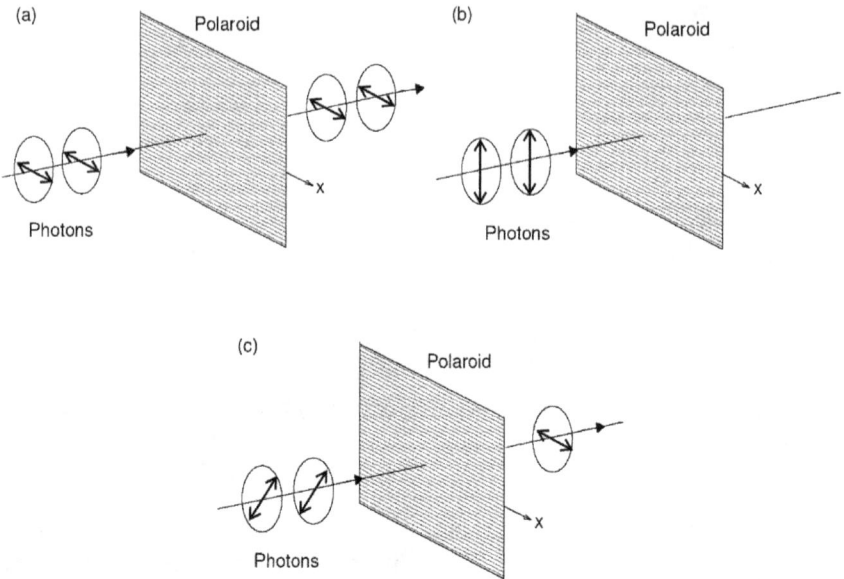

Figure 10.2. Polarised photons incident on polaroid with its axis in the x direction. The photon polarisation is at angle θ to the x-axis. (a) $\theta = 0°$, the chance of a photon being transmitted is 100%; (b) $\theta = 90°$, the chance of a photon being transmitted is zero; and (c) $\theta = 45°$, the chance of a photon being transmitted is 50%. In case (c), any photon which passes becomes polarised parallel to the x-axis.

either transmitted or blocked. If a photon is transmitted, it becomes polarised parallel to the x-axis. (See Fig. 10.2.)

The state of a photon polarised at angle θ to the x-axis may be expressed as a superposition of states polarised parallel and perpendicular to the x-axis. In the notation of §7.3.9, let Ψ_θ be the wave function representing the state polarised at θ to the x-axis, and let Ψ_1 and Ψ_2 be the wave functions representing the states polarised respectively parallel and perpendicular to the x-axis. Then

$$\Psi_\theta = \Psi_1 \cos\theta + \Psi_2 \sin\theta. \tag{10.1}$$

The probability that the photon in the state Ψ_θ is transmitted through polaroid with its axis in the x direction is the square of the coefficient $\cos\theta$ that multiplies Ψ_1 on the right-hand side of Eqn. 10.1. Similarly, the probability that the photon is blocked is the square of the coefficient multiplying Ψ_2. A degree of randomness enters the quantum description of light at this point: it is not possible to predict whether a particular photon will be transmitted through the polaroid or blocked. It is a random process. However if a large number of photons in the same state of polarisation are passed through the polaroid, it will be discovered that the proportion which pass is equal to $\cos^2 \theta$ and the proportion that is blocked is given by $\sin^2 \theta$. The outcome of the measurement on a particular photon is indeterminate before the measurement. When the measurement is made, its wave function collapses to either Ψ_1, if it is transmitted, or Ψ_2 if it is blocked (§7.3.9). Hence, the state of a transmitted photon collapses to one where its polarisation is parallel to the polaroid axis, and the state of a blocked photon collapses to one where its polarisation is perpendicular to the polaroid axis.

But there is a further intuitive difficulty with the quantum description of photon polarisation: in quantum theory, we cannot think of an unpolarised beam of light as consisting of a random mixture of photons each of which has a definite polarisation. Instead, we must treat each photon as unpolarised, which means that it does not have a well-defined polarisation direction. The polarisation of each photon is completely indeterminate, like the position of a particle if its momentum is precisely defined (§9.2). When such a photon is incident on polaroid with axis aligned along the x direction, it has a 50% chance of passing through (in which case its wave function collapses to one describing the state polarised parallel to x) and a 50% chance of being blocked (in which case its wave function collapses to one describing the state polarised perpendicular to x). Of course this would also be true for a polarised photon with well-defined polarisation at 45° to the x-axis, since $\cos^2 45°$ and $\sin^2 45°$ are both equal to 1/2. However in the case of an unpolarised photon, quantum theory predicts that the result is the same *whatever the direction of the polaroid axis*. Whatever direction is chosen for the polaroid axis, there is a 50% chance the photon will pass through, with polarisation identified as parallel to that axis, and 50% chance that it will be blocked and its polarisation identified as perpendicular to that axis. This makes it impossible for the photon to have a well-defined polarisation before it reaches the polaroid, since there is no possible polarisation direction that would be at 45° to all possible directions of the polaroid axis. This indeterminacy is a key ingredient in understanding the EPR experiment with photons.

10.2.2. *An EPR experiment with photons*

In a version of the EPR thought experiment using photons, pairs of photons are produced from a source with their polarisations correlated so that if one has polarisation along any particular direction, then so will the other. However each photon is completely unpolarised, so that if it is incident on a piece of polaroid, the chances of being transmitted and of being blocked are equal for any choice of direction of the polaroid axis.

The two photons in a correlated pair produced by the source fly off in opposite directions, as shown in Fig. 10.3. A detector D_A with its face covered with a piece of polaroid is placed in the path of photon A at a large distance from the source. The polaroid may be rotated so that the direction in space of its preferred axis can be chosen by the experimenter. A similar detector D_B covered with a piece of polaroid is set in the path of photon B. We assume that D_B is further from the source than D_A so that A reaches D_A before B reaches D_B.

Suppose an experimenter at D_A rotates the polaroid so that its axis is along the x direction. Now when photon A arrives, it will either pass the polaroid and register its arrival in the detector (a hit) or it will be blocked by the polaroid and fail to be detected (a miss). In the case of a hit, A's polarisation is measured as parallel to the x-axis, and since A and B were produced with completely correlated polarisations,

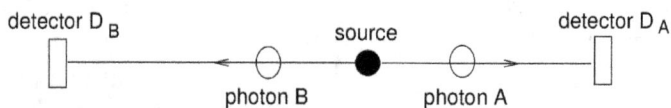

Figure 10.3. Schematic view of the EPR experiment with photons. The source produces correlated pairs of photons, and the two photons in a pair (*A* and *B*) fly off in opposite directions. The face of each of the detectors D_A and D_B is covered with a piece of polaroid.

B immediately acquires a polarisation parallel to the x-axis, no matter how far away it is from the detector D_A. Therefore, when B reaches D_B it will also register a hit, assuming the polaroid axis at D_B is also parallel to the x-axis. On the other hand, if a miss is registered at D_A, a miss will be registered subsequently at D_B.

Imagine two experimenters, one at each of the detectors D_A and D_B which are separated by a large distance. They measure the arrival of a large number of photons emitted in correlated pairs from the source. Each experimenter will observe a random sequence of hits and misses. This is because the individual photons are unpolarised when they leave the source, and each one has a 50% chance of scoring a hit. This probability is completely independent of the results for previous photons. However if both experimenters have set the axes of their pieces of polaroid in the same direction, when they compare results later they will find that their two random sequences of hits and misses match exactly! Whenever a hit was recorded for A, a hit was also recorded for B, and whenever a miss was recorded for A, a miss was also recorded for B. The two random sequences of hits and misses, recorded independently at widely separated detectors, are completely correlated.

What happens if the axis of the polaroid at D_A is set along the x direction, but the axis of the polaroid at D_B is set in a different direction? When A reaches D_A, the polarisation of B immediately becomes the same as that of A: either parallel to the x-axis if A scores a hit, or perpendicular to that axis if A scores a miss. A quantum calculation then predicts, for either polarisation of B, the probability that B scores a hit or a miss at D_B. Again both experimenters observing a sequence of photon pairs will see a random sequence of hits and misses, but now when they compare results they no longer find a complete correlation. The degree of correlation is, however, predicted by quantum theory, and it depends on the angle between the polaroid axes at D_A and D_B. If the polaroid axis at D_A is in direction **a** while that at D_B is in direction **b**, then the degree of correlation is predicted to be

$$E(\mathbf{a}, \mathbf{b}) = \cos(2\theta), \tag{10.2}$$

where θ is the angle between directions **a** and **b**.

10.2.3. *Hidden variables and Bell's Inequality*

Einstein's problem with quantum theory can now be re-expressed in terms of the thought experiment using photons: according to his belief in **local reality**, the measurement on B at D_B should depend only on the settings at this detector and on the properties of B itself, and should not be affected by the measurement on A at the remote detector D_A. Of course Einstein's point of view implies that B must have carried information about its polarisation with it. Einstein concluded that quantum theory is incomplete: if a photon is unpolarised when it leaves the source, it must carry precise values of additional variables which, together with the settings on the detector it enters, uniquely determine the outcome of a polarisation measurement. These **hidden variables** carried by photons A and B are correlated because of the nature of the interaction which produces each photon pair at the source. It is the correlation of the hidden variables that accounts for the correlation between the measurements at D_B and D_A. The nature of the hidden variables is not clear — they cannot simply be values of the photon polarisation, since in this case the predictions of quantum theory concerning the outcome of measurements on unpolarised photons (which have been verified experimentally) would be completely wrong. Their nature would become known as part of a 'complete' theory combining quantum theory with a theory of hidden variables. Bell gave a useful summary of the hidden variables question at a Physics summer school (Bell, 1971).

For most physicists using Quantum Mechanics in the first half century after its development, the problems of hidden variables and the interpretation of the theory were irrelevant. Quantum Theory was providing a new and deep understanding of the structure of materials, of nuclear physics and elementary particles, and of light. Quantum Mechanics gave answers — answers which have completely revolutionised our everyday lives through technological advances such as lasers, computers, superconductors and materials tailored to meet specific needs in industry and spaceflight as well as in the home. Calculations were performed and new insights gained without worrying about whether or not quantum mechanics was incomplete, or what exactly was meant by concepts such as the **collapse of the wave function** and **indeterminacy**. However a few theoreticians still pondered these problems. As early as 1932, the mathematician John von Neumann believed that he had proved that hidden variables could not exist (von Neumann, 1932). Subsequently, fatal flaws were found in his proof, and in 1952 David Bohm produced, as a counterexample, a theory which included hidden variables but reproduced the results of quantum mechanics in the case of the two-slit interference experiment with particles (Bohm, 1952). However Bohm's theory does not satisfy Einstein's locality criterion — in spite of the hidden variables associated with the particles, the interactions have non-local (action-at-a-distance) characteristics.

John Bell became interested in hidden variables in the early 1960s while working at CERN, and his ground-breaking derivation of what is now known as **Bell's Inequality** was published in the first volume of a short-lived journal called *Physics* in 1964 (Bell, 1964). Bell phrased his discussion in terms of Bohm and Aharonov's EPR experiment with pairs of spin half particles, but here his arguments will be translated to the case of pairs of photons which we discussed in the last section. Bell made the assumption that some sort of hidden variables exist, which determine the outcome of a polarisation measurement on a photon. Each photon carries values of these hidden variables with it, and the result of the measurement on a photon depends only on the values of these variables and the setting of the polaroid axis at the detector where it arrives. This implies that the interactions are local, so that what happens at the remote detector D_A cannot directly affect the state of polarisation of photon B. The correlation between the results of measurements on A and B is due entirely to the correlations between the values of the hidden variables carried by the two photons. Based on this assumption, Bell derived a relationship — Bell's Inequality — between the degrees of correlation for different sets of measurements, each set using a different pair of polaroid axis settings at D_A and D_B.

The degree of correlation is defined as the proportion of times the results at D_A and D_B agree (either both hits or both misses) minus the proportion of times the results at D_A and D_B disagree (one is a hit and the other is a miss). If the correlation is complete and all results for a sequence of photon pairs agree, then the degree of correlation is 1, whereas if all results disagree the degree of correlation is -1 (complete anticorrelation). When some results agree and others disagree the degree of correlation lies between -1 and 1, and if there are equal numbers of agreements and disagreements, the degree of correlation is zero.

In 1969, Clauser, Horne, Shimoney and Holt discussed the feasibility of experimental tests of Bell's Inequality using photon pairs (Clauser *et al.*, 1969; Clauser and Shimoney, 1978). They used a version of Bell's Inequality that relates the degrees of correlation for four different sets of measurements:

(i) The polaroid axis at D_A is in a direction **a**, and that at D_B is in a direction **b**.

(ii) The polaroid axis at D_A is in the same direction **a** as in case (i), but that at D_B is in a new direction **b′**.

(iii) The polaroid axis at D_A is in a new direction **a′**, and that at D_B is in the original direction **b**.

(iv) The polaroid axis at D_A is in direction **a′**, and that at D_B is in direction **b′**.

The degree of correlation for the sequence of measurements in case (i) is denoted by $E(\mathbf{a}, \mathbf{b})$, and the degrees of correlation in the other cases are respectively $E(\mathbf{a}, \mathbf{b'})$, $E(\mathbf{a'}, \mathbf{b})$ and $E(\mathbf{a'}, \mathbf{b'})$. The most commonly quoted version of Bell's Inequality reads

$$|E(\mathbf{a}, \mathbf{b}) + E(\mathbf{a}, \mathbf{b'}) + E(\mathbf{a'}, \mathbf{b}) - E(\mathbf{a'}, \mathbf{b'})| \le 2. \qquad (10.3)$$

This expression tells us that the left-hand side is less than or equal to 2. It cannot be greater than 2.

The importance of Bell's Inequality is that the quantum predictions for the left-hand side can give an answer greater than 2 for certain choices of the four directions **a**, **b**, **a′** and **b′**. For example if they are chosen as shown in Fig. 10.4, then the angles between **a** and **b**, **a** and **b′**, and **a′** and **b** are all 22.5°, while the angle between **a′** and **b′** is three

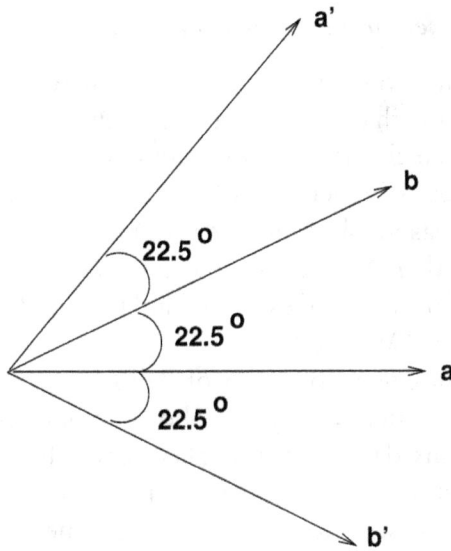

Figure 10.4. Four polariser settings, a, a', b and b', for which the predictions of quantum mechanics satisfy Eqn. 10.5 which violates Bell's Inequality.

times this (67.5°). Using the quantum prediction given in Eqn. 10.2, we find

$$E(\mathbf{a}, \mathbf{b}) = E(\mathbf{a}, \mathbf{b}') = E(\mathbf{a}', \mathbf{b}) = \cos 45° = \frac{1}{\sqrt{2}},$$
$$E(\mathbf{a}', \mathbf{b}') = \cos 135° = -\frac{1}{\sqrt{2}}, \tag{10.4}$$

and putting these values into the left-hand side of Eqn. 10.3 gives

$$|E(\mathbf{a}, \mathbf{b}) + E(\mathbf{a}, \mathbf{b}') + E(\mathbf{a}', \mathbf{b}) - E(\mathbf{a}', \mathbf{b}')| = \frac{4}{\sqrt{2}} = 2.828. \tag{10.5}$$

This result is clearly greater than 2. Hence, for this choice, there is a direct contradiction between quantum theory and Bell's assumption of local hidden variables. An experimental measurement of the degrees of correlation in cases where the quantum prediction violates Bell's Inequality can in principle decide between a theory of local hidden variables and standard quantum theory.

10.2.4. *Experimental tests of Bell's Inequality*

When Bell's work was published in 1964, it was soon realised, to everyone's surprise, that no existing measurements probed this crucial difference between classical and quantum ideas. In fact, setting up an experiment to test whether Bell's Inequality is violated proved very difficult, and a reasonably definitive result was not obtained until the experiments by Alain Aspect and colleagues in the early 1980s. Aspect has given a brief overview of experimental tests of Bell's Inequality up to 1999 in *Nature* (Aspect, 1999).

The first requirement for a test of Bell's Inequality is an efficient source of appropriately correlated photon pairs. Aspect *et al.* used calcium atoms, raised in energy using lasers, which then decay by emitting pairs of photons, each pair being correlated in the way described in §10.2.2. In our description of an EPR experiment with photons, we imagined using a piece of polaroid in front of each detector. In this case, half the photons would be stopped and fail to enter the detector. In practice we need to detect every photon, not just the 'hits', and measure whether the polarisation is parallel or perpendicular to the chosen axis. This is possible with stacks of glass plates called polarisers, which effectively measure whether the photon polarisation is parallel or perpendicular to the polariser axis, and allow all photons to pass through and reach a detector. We shall continue to label the events hits and misses even though the latter do not miss the detector in this arrangement.

The first unambiguous violation of Bell's Inequality was demonstrated by Aspect *et al.* in an experiment which was essentially that described in §10.2.2 (Aspect *et al.*, 1982a). However, this experiment suffered from a fundamental defect: the polariser axes were set at the beginning of the experiment and remained fixed. There is plenty of time for signals travelling no faster than light to pass between the two polarisers, and between the polarisers and the source. Hence, the correlation between measurements on the photons A and B in each pair might be due to correlations between the polarisers and the source, or between the two polarisers. To exclude such effects, the settings of the polarisers should be changed randomly while a photon pair is in flight. Later in 1982, Aspect *et al.* included a random switching mechanism in their experiment, though their arrangement was not perfect since the separation between polarisers A and B was, at 12 m, not great enough to completely rule out normal signals (travelling no faster than light) between the measurements, and it was only possible to switch between two different polariser settings. Nevertheless this was a ground-breaking experiment, which showed a violation of Bell's Inequality (Aspect *et al.*, 1982b). It was not until 1998 that an improved experiment, in which two independent observers were separated by 400 m, ensured that there could be no normal signals between the measurements on the two photons of an entangled pair (Weihs *et al.*, 1998).

Many different experimental tests have been performed since then, all confirming that Bell's Inequality is violated in agreement with quantum predictions. However, no experiment can be perfect. In particular, no detectors or electronic coincidence counters have 100% efficiency, so that not all photon pairs are identified and detected. A bias in the pairs which are successfully identified and measured could, conceivably, cause the results to be skewed towards violation of the inequality. Experiments have been performed using more efficient detectors, increased distances between the two polarisers, and improved random switching of the polariser settings. Advances in non-linear optics have provided new ways to produce correlated pairs of particles, and to perform EPR experiments in which position and momentum rather than polarisation are the measured quantities, as in Einstein's original thought experiment. But all experiments so far have shown a violation of Bell's Inequality and indicated that nature does not behave according to Einstein's view of **local reality**.

10.3. Quantum Information Theory

The theoretical and experimental investigations of entangled states and EPR-type experiments promoted a new view of quantum theory: quantum particles can be thought of as carrying information. In an EPR experiment (such as that with photons, §10.2.2), it appears that there is an instantaneous interaction between the members of a photon pair when the polarisation of one of them is measured. This apparently contradicts the **Special Theory of Relativity**. But notice that the correlation between polarisation measurements of photons A and B cannot be discovered until the experimenters exchange information about their results by normal, classical means. Hence, the degree of correlation cannot be determined instantaneously. Furthermore, it is impossible for an experimenter to use one photon of a correlated pair to send a message to the other experimenter since he has no way of controlling the polarisation of the photon he measures. In whatever direction he sets his polariser, the result of his measurement is completely random. Therefore, although there is an instantaneous correlation between the photons, this does not imply that information can be transmitted at a speed faster than light.

In classical information theory, the basic unit of information is the bit, which takes one of two mutually exclusive values, for example: true or false, hit or miss, 1 or 0. Conventional computers encode all numbers in terms of 1s and 0s (the binary system), and operate on a unique binary input to produce a unique binary output. The values of any physical property with only two possible values may be labelled 1 and 0 and used to generate binary numbers. For example, the polarisation of a photon may be parallel (1) or perpendicular (0) to any chosen axis; or the trajectory of a particle in a two-slit interference experiment may pass through one slit (1) or the other (0).

However, in quantum physics, properties such as the polarisation of a photon or the trajectory of a particle may be indeterminate (§10.2.1). This is given mathematical expression by writing the wave function describing the state of the particle as a linear superposition of wave functions corresponding to all possible results of a measurement. For example, Eqn. 10.1 gives the state of a photon polarised at angle θ to the x-axis as a superposition of the two states parallel and perpendicular to that axis. In this state, the polarisation parallel to the x-axis is indeterminate, and the probability of finding it to be parallel or perpendicular to that axis if a measurement is made is determined by the squares of the coefficients of Ψ_1 and Ψ_2. Both possible results are contained in the state Ψ_θ. A quantum particle can therefore carry information not simply as 1 *or* 0, but as 1 *and* 0. This leads to new ways of transmitting information, and the basic unit of quantum information is called the **qubit**.

10.3.1. *Quantum computing*

Perhaps the most speculative application of quantum information theory is in quantum computing. Richard Feynman (Feynman, 1982), was the first to recognise the potential of quantum computing. Then David Deutsch laid the foundations of the theory of quantum, as opposed to classical, computers (Deutsch, 1985), though in 1985 it was not clear that a quantum computer could ever become a physical reality. Digital computers use logic gates to process binary numbers in a sequence of operations, and each operation processes a single number at a time. A quantum computer, however, could process many numbers in parallel, making such a computer exponentially faster than a conventional one.

A digital computer represents numbers by combining bits: a single bit can be in a state representing either 1 or 0; two bits can represent one of four numbers, and N bits can represent one of 2^N numbers. Table 10.1 shows how each of the eight ($= 2^3$) numbers from 0 to 7 may be represented in the binary system by choosing the values of three bits.

A quantum computer, on the other hand, could represent the eight numbers 0 to 7 simultaneously in an entangled state of three particles,

Table 10.1. Binary representation of the numbers 0 to 7 encoded using 3 bits, each of which can take the value 0 or 1.

Number	bit 1	bit 2	bit 3
0	0	0	0
1	0	0	1
2	0	1	0
3	0	1	1
4	1	0	0
5	1	0	1
6	1	1	0
7	1	1	1

each carrying one qubit of information. An entangled state of N such particles would be able to represent simultaneously 2^N numbers, from 0 to 2^{N-1}. Each operation of the quantum computer could then process all 2^N numbers in parallel.

The physical structure and operation of a quantum computer would be completely different from that of a conventional computer. It must be possible to encode qubits to represent large numbers, and the entangled states must be preserved from collapse through unwanted interactions. Furthermore, the programming of a quantum computer will be entirely different from that of a conventional computer. So far, only two programming algorithms have been found for a quantum computer: the first, Shor's Algorithm (Shor, 1994), factorises a very large number; the second, Grover's Algorithm, published in 1996, sorts items in a list. A good pedagogical review of the second is given by Grover in the *American Journal of Physics* (Grover, 2001).

Shor's Algorithm, if put into practice, would break the security of current transmissions over the internet. These use public key encryption to preserve the security of sensitive data such as credit card numbers. The key used to encrypt the secure message is a product of two very large prime numbers. The product may be public knowledge, but the message itself may only be decoded if the two primes are known. The security of this key depends on the fact that factorising the product using a normal digital computer takes so many steps that it is impractical — the result could not be found within a useful length of time. However, a quantum computer, since it would be able to process many numbers simultaneously, could reduce the time significantly, so that the security of the message would no longer be guaranteed. However, until a quantum computer becomes physically practicable, public key encryption remains secure.

10.3.2. *Quantum cryptography*

Secure transmission of information, which prevents anyone but the intended recipient reading it, has been important since the time of the ancient Greeks. The sender must encrypt the message using an algorithm and then the coded message is sent to the recipient, who will need to know the key to the algorithm which was used in order to decode and read it. However, most codes used in the past can be broken, allowing the secret message to become known to unauthorised people. The public key encryption mentioned in the last section has only become practicable with the advent of modern computing power, and depends on the computational difficulty of factorising the product of very large primes for its security. Breaking this code is impractical rather than theoretically impossible.

However, there is one type of classical code which is unbreakable, at least in theory: the key which defines the encryption algorithm is a string of random numbers (at least as long as the message itself), and a new key is used for each message. The coded message may be sent by normal public channels because nobody who is not in possession of the key can decipher it. However, there are loopholes in this system if classical methods are used for creating and distributing the key. Firstly, it is not easy to generate a truly random string of numbers. The random number generators in computers can only produce strings of pseudo-random numbers, and a given string may be repeated when the same 'seed' is provided to the software. Secondly, there is no way to ensure that the key is not intercepted by an unauthorised user during its transmission from sender to recipient. Thirdly, the sender may not use the scheme correctly, and, for example, might send several messages encrypted using the same key.

Quantum physics provides a way of closing the first two loopholes. Passing a stream of unpolarised photons through a sheet of polaroid or a polariser provides a completely random sequence of hits and misses (§10.2.2). If each hit is labelled by 1 and each miss by 0, the result will be a truly random string of 1s and 0s. The problem of interception may also be avoided because a measurement on a quantum particle interferes with its state, and this can alert the sender and recipient to the presence of an eavesdropper. This use of quantum physics to generate and transmit securely an encryption key is called quantum cryptography.

10.3.3. *Quantum cloning and quantum teleportation*

Quantum cloning means the exact duplication of the state of a quantum system, leaving the original state unchanged. However, the fact that a measurement made on a quantum particle such as a photon changes its state in an unpredictable way means that it is impossible to make an *exact* clone of the state of a quantum particle or system of particles. This is expressed in the 'no cloning theorem' (Wooters and Zurek, 1982). Nevertheless, it is possible to produce an *approximate* clone, and its *fidelity* is a measure of how close it is to the original. See the 2005 review by Scarani *et al.*

Although a quantum state cannot be cloned, '**quantum teleportation**' is possible. This is the exact re-creation, at a distant point in space, of the quantum state of a particle. Because cloning is impossible, this can only occur if the original state is destroyed in the process. Quantum teleportation is achieved through an intermediate state in which three particles are entangled. (Entangled states of three or more particles are of great interest both for theoretical reasons and in the practical development of quantum computers.)

To understand how teleportation works in theory, suppose the unknown polarisation state of a photon P at point A is to be teleported to a distant point B. The original state of photon P cannot be determined since any measurement may alter it. To teleport this state, photon P is first entangled with one of a pair of entangled photons, P_1 say, and since the second photon of that pair, P_2, is entangled with P_1 the result is an entangled state of the three photons, P, P_1 and P_2. Next, P_2 is sent to point B, and a particular type of measurement (called a *Bell-state measurement*) is performed at A on the joint state of P and P_1. (Note that this is not a measurement on the two photons separately, but on the pair as a whole.) This measurement affects the third entangled particle P_2, changing its state to one related to the original state of P by a unitary transformation. Simultaneously, the original state of P is destroyed. Information detailing exactly which Bell-state measurement was used and the result of that measurement is sent to B by normal classical channels, and this determines the relationship between the state of photon P_2 (at B) and the original state of photon P. All that remains is to perform the necessary unitary transformation on P_2 to reproduce the original state of P at the distant point B.

The experimental realisation of quantum teleportation is more difficult to achieve than the previous paragraph may suggest. However, optical systems have been used by various teams to implement Bell-state measurements and the associated unitary transformations on photons. See, for example, experiments performed at the University of Vienna (Bouwmeester *et al.*, 1997; Ursin *et al.*, 2004). Other groups have succeeded in teleporting atomic states (Riebe *et al.*, 2004; Barrett *et al.*, 2004).

Approximate cloning and quantum teleportation are very important for the development of quantum information processing and quantum computing, and much effort is currently devoted to both the theoretical and practical aspects of these fields.

10.4. The Quantum Measurement Problem

The idea of the collapse of the wave function representing the state of a quantum particle, mentioned in §7.3.9 and §10.2.1, has always posed a conceptual problem. The mathematical theory of quantum mechanics, whether it is expressed in terms of Schrödinger's Wave Mechanics or Heisenberg's Matrix Mechanics, provides an equation which determines the way the state of a quantum particle evolves *provided no measurement is made on it*. (Schrödinger's version of the equation is given in Eqn. 7.60.) If the particle is in a state where its wave function is a coherent superposition of two or more wave functions representing states corresponding to different possible values of a measurable variable, then there is no way that the wave function can evolve into just one of its components. Suppose a particle is in a state represented by the wave function of Eqn. 10.1. This is a coherent superposition of Ψ_1 and Ψ_2. Then Eqn. 7.60 determines how Ψ_θ develops with time, and the coefficients of Ψ_1 and Ψ_2 might change with time. However, this equation can never cause Ψ_θ to change completely into Ψ_1 or Ψ_2. Such a change would be irreversible, as there is no way of regenerating the second component of Ψ_θ once the collapse has occurred.

The equation which determines the evolution of the wave function can therefore never describe the collapse to one of its constituent waves. Quantum Mechanics prescribes that if the property associated with Ψ_1 and Ψ_2 (for example, polarisation relative to the x-axis) is measured, then the probability of a collapse to the state Ψ_1 (polarisation parallel to the x-axis) is given by $\cos^2\theta$, and the probability of collapse to Ψ_2 (polarisation perpendicular to the x-axis) is given by $\sin^2\theta$. This is no more than a recipe for extracting predictions for the results of measurements. There is no explanation of how the collapse occurs, or why the normal evolution is circumvented. This problem, of what the collapse of the wave function means in physical terms, whether it actually occurs and if so how, is the quantum measurement problem, which is still a subject of research.

10.4.1. *Schrödinger's Cat and the collapse of the wave function*

Schrödinger was very unhappy with the probabilistic interpretation of his wave function, and in particular with the idea of 'quantum jumps' and the collapse of the wave function. In a paper published in *Die Naturwissenschaften* in 1935, he introduced his famous Cat Paradox (§9.4.2) in the following words [as translated — see Schrödinger, 1935(b)]:

> "One can even set up quite ridiculous cases. A cat is penned up in a steel chamber, along with the following device (which must be secured against direct interference by the cat). In a Geiger counter there is a tiny bit of radioactive substance, *so* small, that *perhaps* in the course of the hour one of the atoms decays, but also, with equal probability, perhaps none. If it happens, the counter tube discharges and through a relay releases a hammer which shatters a small flask of hydrocyanic acid. If one has left this entire system to itself for an hour, one would say that the cat still lives *if* meanwhile no atom has decayed. The psi-function of the entire system would express this by having in it the living and dead cat (pardon the expression) mixed or smeared out in equal parts."

By 'psi-function' Schrödinger means the wave function. The set-up entangles the state of the cat with that of the radioactive atom, and since the state of the atom is indeterminate before it is measured (it is a superposition of decayed and undecayed states) the state of the cat is also indeterminate. The indeterminacy is something we have to accept for quantum particles, but Schrödinger's thought experiment entangles the state of the quantum particle (the atom) with the state of a macroscopic object (the cat). Notice, however, that if this experiment were carried out, it could not prove or disprove the idea that the cat's state becomes indeterminate, since the observed state of the cat will always be either alive or dead and never any combination of the two. Schrödinger's thought experiment merely highlights the problem with measurements on a quantum system: how and at what stage does the wave function collapse?

According to Bohr's Copenhagen Interpretation, classical apparatus is always required to detect and measure quantum systems. There is a dividing line between things which must be treated quantum mechanically and those which must be treated classically. It is not possible, according to this point of view, for classical systems to be described using quantum mechanics, or for a transition from quantum to classical behaviour to be identified from within quantum mechanics. Even worse, Bohr pointed out that where the dividing line is drawn depends on the measurement being made. It is not possible to find a universal criterion to distinguish the level at which a system becomes classical.

The Schrödinger equation (Eqn. 7.60) is symmetric under time inversion, which means that (like Newton's laws) it can determine an earlier state of the particle from its current state. But the process of wave function collapse is irreversible. Once a particular result has been found for the polarisation of a photon, say, the original state which was a superposition of different possible polarisation states is destroyed and lost forever. There is no way of determining from the state of the photon after the measurement what particular combination of states described its polarisation before the measurement. Hence, irreversible wave function collapse is a completely distinct process from the evolution of a quantum state according to the time-dependent Schrödinger equation, and this latter evolution can never result in complete collapse to one component of the wave function with the annihilation of all other components.

Everett's '**Many Worlds**' interpretation of quantum mechanics (Everett, 1957) avoids this difficulty by asserting that the wave function never actually collapses. Instead, the state of the universe as a whole is described by a wave function and each time a measurement is made this wave function splits into different branches, in each of which a different possible result is observed. Since observers are part of the whole system, the observer exists in each of these branches of the universe, but in each one he sees a different result. It appears to the observer in one of these branches that the wave function has collapsed because he is unaware of all the other branches in which his alter ego has made a different observation. In this case, the apparent collapse seems to be associated with the consciousness of the observer. A modern version of the many-worlds interpretation sees the split of the wave function of the universe as due not to observation by a conscious being but to 'decoherence'.

'Decoherence' is the loss of coherence between different components of a wave function which is a linear superposition, such as Ψ_θ in Eqn. 10.1, due to interactions with the environment. For overviews of decoherence, see Zurek (1991) and Bacciagaluppi (2004). Since the 1980s it has been seen as a potential solution to the measurement problem, and the transition from quantum to classical behaviour. Such a solution would extend the use of a quantum mechanical description to everything, rather than assuming a dividing line between quantum mechanical and classical behaviour, and would explain how quantum characteristics are gradually lost through interactions.

Effects which are typical of quantum systems depend on the existence of coherent superpositions of states [of which Eqn. 10.1 is a simple two-state example] which can show interference phenomena (§7.3.9). In practice, the interaction of one particle with others will produce a multi-particle entangled state. For the result of a measurement to be detectable, a macroscopic change such as a pointer movement must occur. To produce this, the interaction of the measured particle with particles in the apparatus must spread to a macroscopically-large number of particles. It becomes extremely difficult to observe interference effects, since this would require monitoring of all the many entangled particles. Even in the absence of deliberate measurement, it is impossible for a particle to remain completely isolated. There will be continual random interactions with the environment, with the result that coherence 'leaks out'. In effect, because we do not monitor the whole environment with which a particle is interacting, the typically quantum interference effects are not detected, and the quantum probabilities for the result of a measurement become indistinguishable from classical probabilities.

There is no question that decoherence effects do occur in practice. In applications of quantum information theory such as quantum computing and quantum cryptography, steps must be taken to preserve the quantum coherence on which they depend. Photon pairs with correlated polarisations rapidly lose coherence, and hence their correlation, if they travel through air. In practice they are transmitted through optical fibre which minimises this loss. However, it is still an open question whether or not decoherence can explain the apparent collapse of the wave function, or provide a way of seamlessly applying quantum theory from the microscopic to the macroscopic level.

In spite of the advances made over the last quarter of a century, the quantum measurement problem is not yet solved. Perhaps a completely new insight will transform our understanding of quantum mechanics during the 21st century. We can only wait and see.

10.5. Further Reading

1. Bell, J. S. 2003.
   Speakable and Unspeakable in Quantum Mechanics, 2nd edn. Cambridge: Cambridge University Press.
2. Rae, A. 2004.
   Quantum Physics: Illusion or Reality? 2nd edn. Cambridge: Cambridge University Press.
3. Singh, S. 1999.
   A quantum leap into the future. In *The Code Book*, London: Fourth Estate, Chap. 8.
4. Wheeler, J. A. and Zurek, W. H. (eds.) 1983.
   Quantum Theory and Measurement Princeton: Princeton University Press.

Chapter 11

A Reflective Interlude

Reality Aspects of Wave–particle Duality

"As far as the laws of mathematics refer to reality, they
are not certain; and as far as they are certain, they do not
refer to reality"   — attributed to Albert Einstein

11.1. Introduction

In Chap. 7 we saw how de Broglie extended Einstein's wave–particle
duality of photons to be applicable to material particles and how
this led, through his own work and that of Schrödinger and others,
to the subject of quantum mechanics as we know it today. However, although it continues, extremely successfully, to explain most
objective data and to underpin the modern technological world,
nobody really understands it. As we have seen in earlier chapters, it
has led to problems concerning the meaning of e.g. 'phase waves',
'pilot waves', 'guidance', 'entanglement', 'non-locality' ('spooky
action at a distance' as Einstein called it), 'hidden variables', 'other
worlds', etc. for which there variously seem to be no explana-
tion/understanding/interpretation in terms consistent with reality.

These topics are further explored in Chap. 12. Meanwhile, this
chapter is a reflective interlude, in essentially descriptive terms, about
the problems of trying to visualise the implications of the developments due to de Broglie, from his original work of 1922–1923 up

to the time of the historic 5th Solvay Conference, held in Brussels in 1927, and its immediate aftermath.

Firstly, it will be useful to have some idea of the importance, or otherwise, that de Broglie himself attached towards these aspects of his work. We only had a glimpse of this in Chap. 7, but his general philosophy on the subject has been analysed and written about in detail, notably by Georges Lochak — de Broglie's biographer and one-time colleague — in his essay 'Louis de Broglie's Conception of Physics' (Lochak, 1993 — see Bibliography).

Naturally, de Broglie's attitude changed as his work progressed, but we can usefully note some general features that formed a background influence.

We are told that he refrained from attributing any 'ontological' value to abstract mathematical representations, especially in multidimensional spaces: they were convenient tools with no physical reality. (**'Ontological'** — 'the nature of being' branch of metaphysics, i.e. what things are actually like, as opposed to **'epistemological'** which refers to *how* we *obtain* knowledge of things.)

On the other hand, he thought that, however provisional, a sufficiently general and experimentally confirmed theory can allow an ontological aspect which can often continue usefully as a limiting case in a new theory. However, to be fully acceptable, any theory whose experimental predictions are correct *must also* provide a picture of the world.

Despite being the architect of wave–particle duality as applied to matter particles, de Broglie put particles first — in contrast to Bohr, the architect of wave–particle complementarity, and opposite to Schrödinger who dispensed with particles. He is quoted by Lochak as saying "it is always the particles that one observes, not the waves". However, although he was aware of the predominance of particles in experimental data, he realised that predictions about particles (matter or photons) can only be based on arguments in terms of waves, i.e. field theories.

Here lay problems that in various ways continue today. An immediate one at that time was de Broglie's difficulty in dealing with optical interference (which we shall refer to later) and another was due to the fact that a law governing the time-evolution of a field does not define the motion of the particles. With regard to the latter, de Broglie tells us (see e.g. de Broglie, 1964(a)) that the work of Einstein, and earlier attempts by (originally) Gustav Mie in 1912, to unite fields and particles in connection with electromagnetic fields, lay behind his proposal in 1924 to treat a particle as "a kind of local accident, a singularity within an extensive wave phenomenon". Another problem here concerns the fact that the equations of quantum mechanics are **linear** (cf. §7.3.7). Though important in the probabilistic interpretation of quantum mechanics, this would not deal with wave-packet stability and the guidance of wave packets along trajectories. That would need waves of greater complexity, i.e. **non-linearity**, and it led de Broglie to seek waves that would guide particles. For this, he saw the need to find what became known as 'hidden variables', or 'hidden parameters', and the need to test them experimentally. He strongly believed that a statistical theory, such as was the current basis of quantum mechanics, is never a complete description of physical reality. Basically a causalist, he was opposed to the **formalism** (i.e. disregard of the meaning of the maths) and **indeterminism** believed in by most of his peers. This is reflected in his book *Non-linear Wave Mechanics, a Causal Interpretation* (de Broglie, 1960 — see Bibliography).

With this brief attempt to form a picture of de Broglie's philosophy — which we shall return to in §11.10 — we can now look at how his theoretical work developed against that background.

11.2. de Broglie: The Beginning

Louis de Broglie's first publications were in 1920 and dealt with the absorption of X-rays in terms of Bohr's atomic theory. Then, with his elder brother, Maurice, in the latter's laboratory, he investigated spectra arising from the photoelectric effect.

Louis continued to contribute on the above topics but, very importantly for what lay ahead, they drew his attention to black-body radiation, and in 1922 he published a paper 'Black-body radiation and light quanta' (de Broglie, 1922(a)). In this, he treated black-body radiation in a new way — solely from the point of view of thermodynamics and kinetic theory, i.e. excluding electromagnetic theory. He considered radiation as a gas of 'light atoms' with energy $E = h\nu$ and chose to neglect 'molecules of light' with energy multiples of $h\nu$. The mass of these 'atoms of light' would be given by $h\nu/c^2$ and their momentum would be $h\nu/c$ (see Eqn. 4.3). Associating them with a rest mass m_0 (*masse propre*) he could also write (cf. Eqn. 2.11)

$$E = h\nu = \frac{m_0 c^2}{\sqrt{1 - \beta^2}} \quad \beta = \frac{\upsilon}{c}, \tag{11.1}$$

where he had to assume that β had a value, less than one, giving the speed υ of the light quantum 'infinitely near' c, and that m_0 was 'infinitely small'. (See §2.3.2 re. use of term 'rest mass'.)

Using kinetic gas theory — significantly without invoking electromagnetic theory — for an assembly of such 'atoms of light' de Broglie then successfully derived the Stefan–Boltzmann Law and Wien's Radiation Law (§3.1) and thought that allowing for polyatomic 'molecules of light' might well lead to Planck's Radiation Law.

In 1922 de Broglie also published a paper on 'Interference and the theory of light quanta' (de Broglie, 1922(b)).

It was at this point that de Broglie began to be led to the notion of **'phase waves'** and to his basic ideas of **'undulatory mechanics'**, and to his **'matter waves'**. The developments leading to these concepts started with the history-making trio of 'Notes' presented in rapid succession in 1923 (de Broglie, 1923(a)(b)(c)).

1923 First Note ('Waves and Quanta')

Without declaring his intentions, de Broglie started by considering a 'material moving object' (*un mobile*) of rest mass m_0 and 'internal energy' given by $E = m_0 c^2$. He said that the 'quantum principle' suggested associating this internal energy with a **'simple periodic phenomenon'** of frequency ν_0 such that

$$E(= m_0 c^2) = h\nu_0. \tag{11.2}$$

Moving at speed $v = \beta c$, $\beta < 1$, with respect to a fixed observer, its energy would then be

$$\frac{m_0 c^2}{\sqrt{1 - \beta^2}} \qquad (11.3)$$

with associated frequency v given by

$$v = \frac{m_0 c^2}{h\sqrt{1 - \beta^2}} = \frac{v_0}{\sqrt{1 - \beta^2}}. \qquad (11.4)$$

On the other hand, again to the fixed observer, the frequency v_0 (Eqn. 11.2) of the internal 'periodic phenomenon' would be reduced to v_1 given by

$$v_1 = v_0\sqrt{1 - \beta^2}. \qquad (11.5)$$

De Broglie described this internal periodic phenomenon as 'varying like'

$$\sin 2\pi v_1 t. \qquad (11.6)$$

At time t the mass would be at a distance $x = vt$ from the origin and its 'internal motion' would then be represented by

$$\sin 2\pi v_1 \frac{x}{v}. \qquad (11.7)$$

De Broglie then 'supposed' that at time $t = 0$ there was also, 'coincident' with the mass, a **'fictitious wave'** moving in the same direction as the mass, with a frequency v_1 the same as that associated with the energy of the mass (i.e. Eqn. 11.4) but (*without explanation*) with speed c/β. This wave, with speed greater than c could not transport energy and was to be regarded purely as a fictitious wave associated with the motion of the mass.

At time t the wave would be represented by

$$\sin 2\pi v \left(t - \frac{x}{c/\beta} \right) = \sin 2\pi v x \left(\frac{1}{v} - \frac{\beta}{c} \right) \qquad (11.8)$$

and he showed that phase agreement ('harmony') was maintained between the 'fictitious wave' defined in this way and the 'internal periodic phenomenon' of the moving object. Equating Eqns. 11.7 and 11.8

$$\sin 2\pi v_1 \frac{x}{v} = \sin 2\pi vx \left(\frac{1}{v} - \frac{\beta}{c} \right)$$

$$= \sin 2\pi vx \left(\frac{1}{v} - \frac{v}{c^2} \right)$$

and hence

$$v_1 = v(1 - \beta^2)$$

which is, as he said, "clearly satisfied by the definitions of v and v_1". (Eqns. 11.4 and 11.5, respectively.)

De Broglie commented that "this important result rests uniquely on the principle of Special Relativity and on the correctness of the quantum relationship as much for the fixed observer as for the moving observer".

He then extended the above features of a material object — its having an associated 'fictitious wave' in phase agreement with an 'internal periodic phenomenon' — to the 'atom of light' which he had shown in his 1922 paper above could be regarded as a moving object of infinitely small mass travelling with a speed infinitely near c.

He next considered the case of an electron in a circular atomic orbit, and showed that the stability of the electron and wave association was in accord with the Bohr–Sommerfeld Theory. We omit the details because they throw no further light on the interpretation of the associated fictitious wave.

The first Note ends with de Broglie stating a belief that it would now be possible "to explain the phenomenon of diffraction and of interference taking into account the quantization of light". They were the subject of his second Note.

1923 Second Note ('Light Quanta, Diffraction and Interference')

So far, de Broglie was regarding a particle, a 'moving body' (*un mobile*), travelling with speed $v = \beta c$, $\beta < 1$, as being accompanied by a sinusoidal 'fictitious wave' travelling with speed c/β (see above), keeping in phase with the 'internal periodic phenomenon' of the moving body.

However, with speed greater than the speed of light the fictitious wave could not transport energy. In this second Note, de Broglie started by saying that one could 'at least' consider the speed of the particle as the '**group speed/velocity**' (*vitesse de groupe*) v_g of the wave motion, meaning there would be a small spread in the values of β. Commenting that the 'physical significance' of the wave would be a task for electromagnetic theory to explain, he chose now to *rename the 'fictitious wave' as a* '**phase wave**' — because its phase at the location of the particle matches the phase of the 'internal oscillation' of the particle.

De Broglie went no further, here, in considering the group velocity aspect but did so later (de Broglie, 1924(b)).

Instead, he moved on to say that the existence of light-atoms requires the modification of Newton's first law (*Principe de L'inertie*). For the dynamics of a freely moving body (*un mobile libre*) he proposed a '**dynamic postulate**', viz. that "at each point of its trajectory it follows the 'ray' of its phase wave" and he sees this in terms of the Principles of Fermat and Maupertuis (to which we refer in §7.3.1 and §12.4.1). This became his '**New Dynamics**'.

Turning to diffraction, de Broglie saw the above as explaining how, if a particle passes through an aperture small compared to the wavelength of its associated phase wave, its trajectory will be curved (*se courbera*) like a diffracted wave. This would explain diffraction of light waves 'however small their number', and (without commenting on changing from light-atoms to material particles) he states that electrons passing through a small enough aperture would show the effect. To that end, he said that an experimental test was necessary. As we know, confirmation did follow in 1927, in the hands of Davisson and Germer (§7.2.1).

De Broglie then conceived of the phase-wave as **guiding** the movement of energy (*déplacements de l'énergie*) "allowing a synthesis of waves and quanta". (In 1927 these guiding phase waves became his '**pilot waves**' (*l'onde pilote*)).

He added that this '**New Dynamics**' is to the old (classical) dynamics as wave optics is to geometric optics (cf. our §7.3.6. Eqn. 7.31).

In conclusion, de Broglie speculated, non-mathematically, about **optical interference** in terms of the crossing of phase waves, and possible coupling effects between light-quanta. This idea was to be superseded in Chap. 5 of his thesis and again in his paper 'On the dynamics of the light quantum and interference' (de Broglie, 1924(d)).

1923 Third Note ('*Quanta, the Kinetic Theory of Gases and Fermat's Principle*')

In this Note, de Broglie considered the statistical aspects of a gas of non-relativistic particles accompanied by phase waves. He asserted that a gas of atoms would only be stable if the waves corresponding to the atoms formed a system of stationary waves. Following Jeans, he considered the stationary modes in a given volume, "transporting zero, one, two or several atoms", with probabilities given by a Boltzmann distribution. This gave him the Maxwell distribution, and for a gas of photons, the Planck distribution — justifying, as was his stated aim, Planck's use of phase space (touched on in §3.2.3 and App. C.4).

Finally, in that third Note, de Broglie enlarged on how his new **dynamic postulate** was in accord with "the fundamental link which unites the two great principles of geometrical optics and of dynamics …".

*　　*　　*

The above historic trio of 'Notes' in 1923 formed the basis of de Broglie's contribution towards the development of Wave mechanics. His doctorate thesis, presented for examination at the Sorbonne in Paris on 25 November 1924, was essentially an elaboration of them.

In 1923, he had also had a short note 'Waves and Quanta' published in the journal *Nature* (de Broglie, 1923(d)), and early in 1924 a paper 'A tentative theory of light quanta' in *Philosophical Magazine* (de Broglie, 1924(a)). In these, he was presenting, for the first time in English, an outline summary of the above ideas.

Just before his thesis examination, de Broglie had three further Notes (de Broglie, 1924(b)(c)(d)) communicated to the Academy of Sciences in Paris in 1924, with titles as follows (English translation):

1924(b) 'On the general definition of the correspondence between wave and [particle] motion'. It was here that he established the identification of particle speed with the **group velocity** of the phase wave referred to in his second Note of 1923 above. For this he used a Hamiltonian equation of motion, whereas the method we describe in §7.2.1 was that given by him later (de Broglie, 1930 — see Bibliography).

1924(c) 'On Bohr's theorem', in which he showed that the correspondence between atomic transition frequencies and classical radiation frequencies could be deduced from the dynamics of the phase waves connected with the orbital electrons.

1924(d) 'On the dynamics of the light quantum and interference' in which he claimed to have produced an improvement on his account of optical interference in the second of the trio of Notes the previous year and on the ideas that he had already written in Chap. 5 of his thesis. He now reasoned, purely qualitatively, that interference patterns could be explained just in terms of the propagation of the phase waves. The mechanism involved the interference of phase-waves affecting photon trajectories and causing them to bunch together in regions of bright fringes, and it failed to deal with the relationship between phase/guiding waves and light waves (electromagnetic radiation).

However, a proper explanation of optical interference was to elude de Broglie.

* * *

As mentioned in Chap. 7, de Broglie's work was not widely noticed until the publication of his thesis early in 1925, with the same title, in *Annales de Physique* (de Broglie, 1925) to which we now turn.

11.3. de Broglie's Thesis (1924(e)) ('Researches on the theory of quanta')

De Broglie's doctorate thesis examination was on 25 November 1924 (de Broglie, 1924(e)), then published in *Annales de Physique* early in

1925 (de Broglie, 1925). (English translation by A. F. Kracklauer: *On the Theory of Quanta* (2004). Fondation Louis de Broglie, Paris).

It was an elaboration, in seven chapters and an appendix, of his work described above, and a number of points can usefully be recapitulated here from the point of view of his conception of what was involved in the ideas he was developing.

In Chap. 1 ('The Phase Wave'), after an historical introduction, de Broglie discussed the phase-wave concept as he had in the 1923 Notes, and the need for it to be the group velocity that matches the particle velocity (but the implication of a narrow spread of frequencies was not considered until later). Without more detail about this he then considered phase-waves in a Minkowskian context.

Chap. 2 ('The Principles of Maupertuis and Fermat') consisted of a lengthy discussion of the Fermat–Maupertuis parallel, supported by his hypothesis of a phase wave connected with all material particles. He showed that the phase of the 'guiding' wave determines the particle velocity (§12.4.1). This 'new law' was to be his **'First-order Dynamics'**. Here he also reiterated his starting point as having been his conviction, stemming from the work of Einstein and Planck, about "the impossibility of considering an isolated quantity of energy without associating a certain frequency with it via the 'quantum relation' $E = h\nu$".

As we saw above, that led him to propose that to each energy packet of rest mass m_0 there is attached a **periodic phenomenon** of frequency ν_0 such that $h\nu_0 = m_0 c^2$ (as in Eqn. 11.2 above) "ν_0 being measured in the rest frame of the energy packet".

In this chapter, de Broglie considered three examples. One was the case of an electron, charge e, with velocity v in an electrostatic potential ϕ. He obtained the following expressions for the frequency, ν, and phase velocity, v_{ph}, of the phase wave:

$$v = (mc^2 + e\phi)/h, \quad v_{\text{ph}} = (mc^2 + e\phi)/mv$$
$$\text{where} \quad m = m_0/\sqrt{1 - v^2/c^2} \tag{11.9}$$

It was these that Schrödinger came to, for the starting point of the development of his 'Wave Mechanics' (§11.4).

Also in this chapter, importantly for our view of his model of the phase wave, he now speculated that the 'periodic phenomenon' is not localised within the energy fragment but is *'spread over an extensive region of space'*.

Another point of special interest, which de Broglie dealt with briefly in his first Note, is in his Chap. 3 ('The quantum conditions for the stability of orbits'). Here, he elaborated further on the question of electrons in atomic orbits and showed, by using the Fermat–Maupertuis parallel and imposing the idea of resonance on the phase wave of an orbiting electron, he could arrive at the Bohr quantum condition (cf. Eqn. 5.17).

We can again acknowledge the great use de Broglie made of the Fermat–Maupertuis parallel for which he deserved more credit than accorded him in our Chap. 7 here. The fact that, as shown in §7.3.6, division by h connects the two, was commented on by de Broglie — in the context of the resonance condition being identified with the stability condition from quantum theory — as a 'beautiful result' constituting "the best justification that we can give for our attack on the problem of interpreting quanta". As Bacciagaluppi and Valentini put it — "de Broglie had achieved a concrete realisation of his initial intuition that quantisation conditions for atomic energy levels could arise from the properties of waves".

However, de Broglie quickly added that although he had shown "why certain orbits are stable", a theory for the passage from one orbit to another would require a "modified version of electrodynamics, which so far we do not have".

In Chap. 4 de Broglie dealt with 'Motion quantisation with two charges'. For the hydrogen atom he recognised the problem of how to deal with the distribution of the electrostatic energy between the nucleus and the electron, and how to assign masses. Neglecting the question of mass correction, he applied his resonance condition to the relativistic hydrogen atom and obtained an expression that, in the non-relativistic limit, agreed with Bohr's earlier treatment (Bohr, 1913). It is interesting to note that, by assigning phase waves separately to nucleus and electron, he showed that "Bohr's conditions may be interpreted as resonance expressions for the relevant waves. Stability conditions for nuclear and electron motion considered separately are compatible because they are identical".

In de Broglie's Chap. 5 ('Light Quanta') a point of interest for us here was his wanting "to specify more exactly just how one is to imagine an 'atom of light'". He said that for a fixed observer "it appears as a little region of space within which energy is highly concentrated and forms an undividable unit", and has — in contrast to spherically symmetrical electrons — "an axis of symmetry corresponding to the polarisation". The latter caused him to represent a quantum of light as having "the same symmetry as an electrodynamic doublet" — a matter awaiting "serious modifications to electrodynamics".

On the question of optical interference, de Broglie speculated that since, in electromagnetic theory, photoelectric effects involving e.g. absorption are in proportion to field intensity, one could 'imagine' that "where photons traverse an interference region, they can be absorbed in some places and not in others. This is in principle a very qualitative explanation of interference, while taking the discontinuous feature of light energy into account".

However, as we noted above, this was superseded by his comments in his 1924(d) paper and, even then, a proper explanation of interferences was to continue to elude him.

With regard to his having worryingly attributed a rest-mass to the light-atom, albeit it extremely small, he concluded — by using an expression for its propagation velocity and the known measurements of the velocity of propagation of radio waves with wavelengths of several kilometres — that the mass needed to be less than 10^{-44} gm. He returned to the problem in an appendix at the end of his thesis where, after finding further difficulties, he finally acknowledged that "the real structure of radiant energy still remains very mysterious".

In Chap. 6 ('X-ray and γ-ray Scattering'), which concerned the interactions of X-rays and γ-rays with matter, de Broglie derived afresh some details of the Compton Effect.

In Chap. 7 ('Quantum Statistical Mechanics') he expanded, in detail, on his third Note of 1923, concerning his introduction of phase waves into the theory of black-body radiation. Of interest from the conceptual/visual point of view, is his reasoning (in Part III of the chapter) that if several 'atoms of light' have phase waves that superpose, one could say that they are transported by the same wave, and their motions could not be considered as entirely independent in calculating the probabilities: this gives rise to '*une sorte de coherence*' of their motions.

He noted that in view of the apparent rôle of the stationary wave in his treatment of black-body radiation (de Broglie, 1922(a)) he had to introduce the concept of an '**elementary stationary phase-wave**'. This could be defined as a superposition of two identical phase-waves travelling in opposite directions. Each elementary stationary phase-wave could carry 0, 1, 2 ... 'light atoms', and following this through in the familiar Boltzmann way led him to Planck's radiation formula, and also to Einstein's expression for the energy fluctuation of black-body radiation (cf. §4.4).

11.4. A Brief Stocktaking

De Broglie had succeeded in explaining a wide variety of phenomena with his phase-wave model. Our aim has been to note the aspects that followed the visualisation of the model involved, rather than look at the theoretical details that are to be found in the original papers.

We have seen that his achievement was based on adopting Einstein's light-quantum hypothesis and the setting up of a unification of a dynamical description of light-quanta with the dynamics of atoms and electrons. For this, he had needed to endow the light-quantum with an infinitesimally small mass and to treat it as a particle moving with relativistic speed. Whilst Einstein had deduced a particle property for light in his explanation of the photoelectric effect, de Broglie was now turning this around and associating all particles of matter (now to include light quanta) with a wave motion.

With his phase-wave concept de Broglie formulated a 'New Dynamics', a quantum-theoretical dynamics describing both matter particles and light-quanta. In his second Note of 1923 he added to the phase-wave the property of it '**guiding**' the movement of energy (*déplacements de l'énergie*) and 'allowing a synthesis of waves and quanta', but he did not elaborate.

An important point to note here is that de Broglie's New Dynamics was **a first-order (velocity-based) dynamics** (cf. Bohm's acceleration-based wave mechanics).

Although the use of the phase-wave concept was proving very successful, he was very aware that, as we have seen, it left much that was vague. The meaning/interpretation of 'group velocity' was unclear, with its requirement of a spectral spread. (Much later, looking back at his 1924 work, he said — interestingly — that if the complex wave due to spectral width is represented by a Fourier integral, the Fourier components "only exist in the theoretician's mind", and "**the superposition is the physical reality**" (de Broglie, 1972).)

Also, his treatment of interference (see 1924(d) above) left him unsure about the relationship between his phase/guiding waves and light waves (electromagnetic radiation).

In his thesis de Broglie ended by saying that "The present theory should therefore be considered rather as a scheme whose physical content is not fully defined, than as a consistent doctrine which is definitely established".

We note at this point that de Broglie's phase waves were real-valued functions of space and time with oscillating amplitude and phase. It was Schrödinger who set up the familiar '**wave equation**'.

We now move on to the aftermath of the birth of de Broglie's revolutionary ideas, and see how they developed, led to and interacted with the arrival of Schrödinger on the scene.

11.5. 1924: Schrödinger

In December 1924 Einstein wrote his often-quoted observation to H. Lorentz that he was impressed by de Broglie's attempt to interpret the Bohr–Sommerfeld quantisation rules — "the first feeble ray of light to illuminate the worst of our physical riddles". His support of de Broglie's ideas then materialised when he advocated de Broglie's ideas and showed in the second of his two papers (Einstein, 1925(b)) on the quantum theory of the ideal gas that two distinct terms could be interpreted, one wave-like and the other particle-like (§7.3.2) — duality he had previously already suggested in the context of his 1909 work on black-body radiation (§4.4.1 Eqn. 4.31).

A year later, in November 1925, Schrödinger expressed to Einstein his interest in de Broglie's thesis and whose work he had also read about in Einstein's recent gas theory papers referred to above. It was at this point that Schrödinger embarked on seeking a wave equation for de Broglie's phase waves, and we have noted above that the second chapter of de Broglie's doctorate thesis was his starting point, applying Eqns. 11.9 to the hydrogen atom with a Coulomb field. In a four-part paper (Schrödinger, 1926(a)) he developed his time-independent equation (Eqn. 7.49). (Mehra and Rechenberg (eds.) Vol. 5 Part 2 — see Bibliography.)

The formalism of Schrödinger's wave equation and its further development was to be a more powerful way of tackling problems than de Broglie's way of imposing the idea of phase-wave resonance — as referred to in the third chapter of his thesis.

In §7.3 we outline how Schrödinger's 'Wave Mechanics' was developed. Suffice it to note here that although he had developed his wave mechanics from de Broglie's ideas he had not accepted de Broglie's concept of a particle being localised within an extended associated wave. Instead, he disposed of particle trajectories and worked only in terms of waves.

11.6. Post-1924: de Broglie

Following Schrödinger's arrival on the scene, de Broglie started — seemingly for the first time (in print) — to consider explicit equations for his waves, based on standard relativistic wave equations, but still based on a particle as a very small region of very large wave-amplitude in an extended wave — in line with his group velocity/wave packet concept of 1923. Also, he would require adherence of the particles/singularities to his 'New Dynamics' embracing the Fermat–Maupertuis parallel.

This led to de Broglie's major paper of May 1927 (de Broglie, 1927b), but he first published two Notes in *Comptes Rendus* (de Broglie, 1926(a), 1927(a)) that would contribute to the main paper, and that start to tell us his reaction to the Schrödinger/Born probabilistic approaches. We can note points of interest to us here.

In the first Note (August 1926) he treated **photons** as 'moving singularities', which, he reminded the reader, "constitute the quanta of radiative energy". Adapting the usual classical wave equation, and with Born's probabilistic approach, he deduced (not entirely satisfactorily) that the light-quanta density is proportional to the classical wave intensity of the radiation.

In the second Note (January 1927) de Broglie applied the above approach to the use of the Schrödinger wave equation in connection with the motion of **material particles,** with the probability density here "as envisaged by Born" (i.e. amplitude squared). He ended by declaring that in "micromechanics as in optics" continuous solutions of the 'equations of propagation' can only provide statistical information: an exact microscopic description would surely necessitate 'singularity solutions' showing (*traduisant*) the discrete structure of matter and radiation.

11.7. de Broglie, May 1927 ('Wave Mechanics and the Atomic Structure of Matter and of Radiation')

This 17-page paper (de Broglie, 1927(b)) is generally regarded as de Broglie's most important paper, a detailed exposition of his work up to then. It was here that he proposed his **principle of the double solution,** and the idea of a **physically-real 'pilot wave'** guiding the motion of a particle, concepts that were to play an important rôle in later debate and developments (Chap. 12).

Here, we pick out points that emerge and relate to the development of the 'visualisation' aspect.

De Broglie was now regarding a particle as **localised within an extended wave,** and he was considering specific equations for his waves, using relativistic wave equations, but unlike Schrödinger, with particles as small singular regions of large amplitude within an extended wave. Furthermore, the Fermat–Maupertuis equivalence had to be adhered to in order to unify the dynamics of particles with classical wave-theory, in conformity with his 'New Dynamics'.

He started with a material point-mass, moving in the z-direction in free space with a constant velocity v. The mechanical wave to which he likened this was represented by a solution to an equation of the form he expressed (with reference to de Broglie, 1926(b)) as:

$$\nabla^2 u - \frac{1}{c^2}\frac{\partial^2 u}{\partial t^2} = \frac{4\pi^2 \nu_0^2}{c^2} u \qquad (11.10)$$

where $\nu_0 = \dfrac{m_0 c^2}{h}$ as hitherto (Eqn. 11.2).

[He based the above equation on an early (abortive) attempt by Schrödinger to establish a satisfactory relativistic wave equation, and refined by Oscar Klein and Walter Gordon in 1926 — the year before this paper by de Broglie. ('Klein–Gordon' equation §7.4.).]

Following the reasoning in his First Note and in his thesis, de Broglie considered solutions of the form

$$u(x,y,z,t) = f(x,y,z,t)\cos\frac{2\pi\nu_0}{\sqrt{1-\beta^2}}\left[t - \frac{\beta z}{c} + \tau\right] \qquad (11.11)$$

where τ is a constant.

To represent the **particle**, de Broglie regarded f as a singularity at $z = vt$, i.e. travelling with the wave. Putting

$$\nu = \frac{\nu_0}{\sqrt{1-\beta^2}} = \frac{W}{h} \qquad (11.12)$$

and **wave speed** $V = c/\beta$ (as he postulated for his 'fictitious wave') where W is the particle's internal energy $m_0 c^2$, gave

$$u(x,y,z,t) = f(x,y,z,t)\cos 2\pi\nu\left[t - \frac{z}{V} + \tau\right]. \qquad (11.13)$$

He added that, on the other hand, Eqn. 11.13 showed that the particle could also be 'likened' to a wave group of narrow frequency range.

De Broglie presently considered the "representation of a cloud of points by a continuous wave" not subjected to external or mutual forces, travelling in the z-direction with the same velocity (here v) and with singular amplitudes (representing particles). This 'global phenomenon' could then be represented as:

$$U(x,y,z,t) = \sum_i f_i(x,y,z - vt,t)\cos 2\pi\nu\left[t - \frac{vz}{c^2} + \tau_i\right]. \qquad (11.14)$$

For simplicity, putting τ_i all zero so that the 'material points have the same phase', Eqn. 11.10 'allows the continuous solution':

$$\Psi(x, y, z, t) = a \cos 2\pi \nu \left[t - \frac{\nu z}{c^2} \right]. \qquad (11.15)$$

In a footnote, he noted that Ψ would now refer to **continuous solutions** of the equations of propagation identical to those of Schrödinger.

In the course of this, and with it ever in mind, de Broglie showed that his treatment was in accord with the Fermat–Maupertuis equivalence.

With further detailed theorising and some assumptions along the way, de Broglie then arrived at the conclusion that particle and wave aspects were two solutions of the **same** equation — one being the **point-like singularity** (his '*u* wave') and the other a continuous amplitude wave (his 'Ψ wave'). He described it as his '**principle of the double solution**'. Only valid in free space, however, he said that it remained just a hypothesis in the general case.

Here — though with the above limitation — was an explicit statement of de Broglie's concept, originally mooted in his first Note of 1923, of 'phase harmony' between the internal periodic phenomenon and the accompanying 'fictitious wave' ('phase wave').

From the point of view of visualising what these developments amounted to, the most noteworthy is that he stated (on the final page of his paper) that "... it is necessary to preserve ... the notion of the atomicity of matter...". He went on to add that if one doesn't wish to invoke the principle of the double solution (presumably because of its hypothetical nature in the general case) it was "admissible to recognise, so far as regards distinct realities, the existence of the material point and the continuous wave represented by the function Ψ, and one will take it as a postulate that the motion of the point is determined as a function of the phase of the wave [Eqn. I in Sect.V, §11]". He concluded that "One then conceives of the continuous wave as guiding the motion of the particle. It is a **pilot wave**."

Here was the first mention of the 'pilot wave' that was to become of such interest (Chap. 12). However, he regarded it as a '**provisional theory**', there being a need "...without doubt, to 'reincorporate' the particle into the wave-like phenomenon..." One is left unsure, then, as to how he regarded his 'pilot waves' as guiding the motion of the particle.

There are two final points of interest to us here. One concerns light, and de Broglie showed that, as 'classically', photon density is proportional to the square of the amplitude of the guiding wave. The other was that, as with Schrödinger, for an ensemble of hydrogen atoms in a given state, the mean electronic charge density is proportional to $\Psi\bar{\Psi}$.

11.8. September 1927: Como (Volta Centenary Meeting)

We have seen (in Chap. 9) that running parallel with de Broglie's work there was the arrival of Bohr's idea of '**complementarity**' (initially to deal with wave–particle duality), Heisenberg's Uncertainty Principle (which he found to be a consequence of his matrix mechanics), Bohr's lecture in Como in September 1927, and the emergence of the 'Copenhagen Interpretation'.

There is evidence (Bacciagaluppi and Valentini, 2009 — see Bibliography) that on seeing de Broglie's May 1927 paper, Pauli commented in a letter to Bohr on 6 August that Bohr should refer to it in his Como lecture. In an unpublished version of Bohr's lecture, he did apparently take issue with de Broglie's ideas, claiming them to be unsuited to solving the current problems. However, as we know, Bohr's main purpose at Como (at which de Broglie, Einstein and Schrödinger, were in any case all absent) was to develop his ideas about complementarity.

In the following month, in October, there was the historic 5th Solvay Conference, in Brussels, at which all the leading figures were present and at which de Broglie made an important contribution with his report entitled '**The new dynamics of quanta**'. Together with his May 1927 paper, this saw the birth of **pilot-wave theory** that we associate with his name today.

11.9. October 1927: 5th Solvay Conference

In Chap. 7 we noted that the historic Solvay Conference in Brussels in October 1927 was the first time that Einstein, Schrödinger and de Broglie heard of the deliberations at the Como meeting, with its emphasis on the probabilistic/uncertainty approach to quantum mechanics, and the emergence of the 'Copenhagen Interpretation', to which they were opposed — though it will be recalled (§7.3.9) that Einstein had a predilection for Born's statistical interpretation of Schrödinger's wave mechanics. Opposition was mainly based on quantum mechanics failing to be consistent with reality, failing to deterministically allow prediction of events from initial causes, and failing to 'allow' interactions to be 'local'.

Unfortunately, the six-day Solvay Conference, for which the topic that year was 'Electrons and Photons', was inadequately and unevenly reported, but it has now been extensively investigated and analysed and as complete an account as possible has been assembled (Bacciagaluppi and Valentini, 2009 — see Bibliography).

Apart from 'Reports' by W. L. Bragg, A. H. Compton, and numerous informal discussions, there were reports by de Broglie, Born jointly with Heisenberg, Schrödinger and lastly, remarks by Dirac, and further discussions. Einstein had withdrawn from contributing, apparently because he had recently proposed, in an unpublished lecture to the Prussian Academy of Sciences in Berlin in May, an alternative version of pilot-wave theory in an attempt to counter the incompleteness of quantum theory, but which he abandoned. And Bohr was not invited to contribute, although he requested that a translation of his Como lecture should be included in the published proceedings.

De Broglie's extensive report ('The new dynamics of quanta') included much of the content of his recent *Journal de Physique* (1927(b)) paper, with its 'provisional theory' but now — importantly — its extension to non-relativistic many-body systems (usually credited to Bohm). With this extension, he now presented his 'pilot wave' theory in the form in which it is known today: for a non-relativistic many-body system a guiding wave in configuration space determines the particle velocities according to de Broglie's New Dynamics.

He had disposed of his original concept of a point-like singularity (his u waves) and his Principle of the Double Solution (though he was to return to this later and continue to work on it for many years

(§11.10)): they had served their purpose — as so often happens in the history of physics. He was now dealing with the relationship between particles and Ψ waves.

He began by considering a single relativistic particle in an external electromagnetic field and obtained an expression that classically, according to the Fermat–Maupertuis parallel, gave an expression for the 'velocity field' of the particle, namely the 'guidance equation' (see §12.4.1). He again proposed that this is also valid outside the classical limit and which completely determines the motion of the particle, given its starting position. As in the second chapter of his thesis, he was still proposing a **first-order theory of motion based on velocities**.

For an ensemble of particles guided by the above 'velocity field' he then showed that, ignoring initial positions, $\Psi\bar{\Psi}$ gives the probability for the presence of a particle in a given element of space.

He therefore concluded that "The Ψ wave then appears as **both a 'pilot wave and a probability wave'**".

We are left with a continuing, changing picture concerning reality, and it is not surprising that de Broglie here reasserted his belief that there were no grounds for abandoning determinism (in which respect he differed from Born).

Other topics in de Broglie's report included an outline of Schrödinger's work and a criticism of it. In Schrödinger's many-body system with a Ψ wave propagating in configurational space, but with particles existing only as localised wave packets, de Broglie found it 'paradoxical' that it should be possible to construct a configurational space with the coordinates of points that do not exist.

His report ended with a discussion of recent experiments involving diffraction, interference and electron scattering, and their interpretation. However, a satisfactory explanation of interference had eluded him and it was Born and Heisenberg, in their combined report at the Solvay Conference, who investigated the problem more constructively. Their report was a major contribution to the statistical interpretation of quantum mechanics and now, for the first time, included a treatment of interference — which we may note was possible "*precisely* through the absence of any 'hidden' values for energy" (Bacciagaluppi and Crull, 2009 — see Bibliography; Bacciagaluppi and Valentini, 2009 — see Bibliography).

Of the lengthy discussion of de Broglie's report, the main criticism was from Pauli who found difficulty with de Broglie's theory in the case of inelastic collisions.

De Broglie's work subsequently failed to receive the attention it deserved, partly perhaps, because at the Solvay Conference there was no consensus. Furthermore, the extensive discussions of de Broglie's report were not adequately reported in the published proceedings of the conference.

With his theory, particles of any sort are treated like point-like objects with 'pilot waves' guiding them, and this was seen as what was later to be called a **'hidden variables theory'** (term introduced by Bohm in 1952). Also, neither de Broglie's nor Schrödinger's theories were accepted by Born and Heisenberg, for whom reality is not independent of the observer (§9.2). Furthermore, the main interest at the time was in the recently emerging **Copenhagen Interpretation** concerning the inter-relation of wave–particle duality and Heisenberg's **Uncertainty Principle** (Chap. 9).

Though satisfactorily answering most of the criticisms of his theory, de Broglie was troubled by his failure to deal with the effect of the intervention involved in measurement making. His theory became neglected until it was resurrected in 1952 by Bohm who showed how to use it in dealing with the **'measurement problem'**. Bohm also noted that the theory was **'non-local'**, and as we saw in Chap. 10 this attracted the attention of J. S. Bell in 1964.

The re-examination of the proceedings of the 1927 Solvay Conference referred to earlier, and the aftermath of the work of Bell and Bohm (Chap. 10), and the other related topics mentioned earlier, are among the topics of Chap. 12.

11.10. de Broglie: Post-1927 Solvay Conference

An interesting summary picture of the outcome and the immediate aftermath of the Solvay Conference for de Broglie has been painted by Lochak (Evans and Thorndike, 2007 — see Bibliography).

Two groups had collided at the conference. One included Einstein, Planck, de Broglie, and Schrödinger: they defended a **'causal**

and descriptive physics'. The other was the Copenhagen School (in homage to Bohr), and included Heisenberg, Pauli, (less clearly) Dirac, as well as Bohr himself: they defended a **'formal and indeterministic theory'** where quantum mechanics was seen as a finished and definitive whole — and in Brussels they triumphed.

As de Broglie was leaving Brussels, Einstein encouraged him to continue to search "for a representation of particles as singularities of the wave". However, de Broglie thought that in persisting and probably failing he ran the risk of marginalising himself and so, against his will, he "rejoined the School of Copenhagen". [In his book (de Broglie, 1930) he detailed the three reasons why he reluctantly abandoned his theory, and they were not connected to Pauli's criticism in Brussels.]

In the 1950s he turned away from the Copenhagen School, to return to his unfinished work, on which he published extensively for the next 20 or so years.

The failure of de Broglie's and Schrödinger's theories had been largely due to their equations being linear. Though vital in probabilistic interpretations this did not provide a way for dealing with stabilisation of wave packets (whether or not containing particles as such) or their guidance. As Lochak points out, non-linearity is not a property but the lack of a property, and it becomes necessary to find the right property without a basis for doing so (Lochak, 1992 — see Bibliography).

Accepting the need for **'hidden variables'** as evident, and ever a causalist, de Broglie's interest was now in finding the correct ones and hoping to put them to experimental test. For this, he returned to his theory of the Double Solution, which did define simple hidden variables in the form of particle coordinates.

In 1972 he published a 22-page paper, reissued in 1987 in an English translation because of its historical value, in *Annales de la Fondation Louis de Broglie*, where the editor stated that the intention was to "convey the precise physical meaning, and most importantly, the spirit of Louis de Broglie's work".

Under the title 'Interpretation of quantum mechanics by the double solution theory', de Broglie said that since 1923–1924 he had been "looking for a truly concrete physical image, valid for all particles, of the wave and particle coexistence discovered by Albert Einstein in his Theory of light quanta", adding that he had "no doubt whatsoever about the physical reality of waves and particles". With regard to the 'double solution' theory he said he [still] wished "to insist on the two main and basic ideas of this interpretation of Wave Mechanics". It is interesting to see, in his own words (albeit in translation), how, after 50 years and all that had happened during that time, he was now describing waves and particles.

Firstly, he said that the physical wave, (denoted in this paper as v), has "a very small amplitude ... which is distinct from the Ψ wave. The latter ... has a statistical significance in the usual quantum mechanical formalism". He 'connected' the two by "the relation $\Psi = Cv$ where C is a normalising factor". Then he added that the "Ψ wave has the nature of a subjective probability representation formed by means of the objective v wave". He further added that this distinction was the reason for naming the theory as the '**double solution theory**' since "v and Ψ are thus the two solutions of the same wave equation".

Secondly, he said that for him "the particle, precisely located in space at every instant, forms on the v wave a small region of high energy concentration, which may be likened in a first approximation to a moving singularity".

He stated that in the general case, a "particle's internal vibration is constantly in phase with the wave on which it is carried ... And that this result ... can be considered the main point of the guidance law". Later, he showed that a "characteristic of the guided motion" is that it is "performed according to relativistic dynamics of a variable proper mass".

With regard to Schrödinger's Ψ wave he briefly explained why it could not be "considered as a physical wave", though "its generalisation did lead to accurate prediction and fruitful theories". He added that "the situation is clarified by introducing together with the statistical Ψ wave, the v wave, which being an objective physical reality, may give rise to phenomena the statistical aspect of which is given by the Ψ wave".

Saying that it "becomes important to establish the relationship between the Ψ and v waves", he referred to recent work he had done on this with Andrade e Silva (de Broglie, 1969).

On the question of the "localisation of the particle in the wave and the guidance law", de Broglie regards it as "premature to try to describe the internal structure of the singular region, i.e. the particle ... will probably involve complicated non-linear equations". For the justification of the '**guidance law**' he referred back to his earlier work (de Broglie, 1927(b), 1956).

An interesting, brief section followed, entitled 'The hidden thermodynamics of particles', a topic he had developed as an extension of the 'double solution' theory (de Broglie, 1964(b), 1968(b)). He showed how the energy of a particle, with the original association of its internal 'simple periodic phenomenon' (here a 'small clock') and an internal energy (Eqn. 11.2) that doesn't contribute to momentum, is "similar to that of a heat-containing body in an internal state of equilibrium". He showed that relativistic formulae for a clock's frequency and for heat, does "make the double aspect possible". Following that, he briefly noted its extension to 'the relation between action and entropy'.

The final section has the title 'On the necessary introduction of a random element in the double solution theory. The hidden thermostat and the Brownian motion of the particle and the wave'. This acknowledged that the assumption that a "particle's motion in its wave" cannot be wholly correct, and he reasoned that "a particle's motion is the combination of a regular motion defined by the guidance formula, with a random motion of Brownian character". The case of an electron in a hydrogen atom posed a different problem and he referred to the 'sub-quantum medium' concept of Bohm and Vigier in 1954.

De Broglie concluded this paper by saying that he thought that when the present state of the Wave Mechanics interpretation by the double solution theory, and its thermodynamical extension, is further elaborated and modified in some respects, "it will lead to a better understanding of the true coexistence of waves and particles about which actual Quantum Mechanics only gives statistical information, often correct, but in my opinion incomplete".

Chapter 12

Interpretations of Quantum Mechanics

Sara M. McMurry

12.1. Attempts to Make Sense of Quantum Physics

The mathematical theory of Quantum Mechanics, including relativistic Quantum Field Theory, is one of the most successful theories of all time. It has given us an understanding of the structure and behaviour of matter at the atomic and sub-atomic scales, led to advances in optics and electromagnetism which have changed our daily lives, and contributes to our understanding of the origin of the universe. But it is very hard to understand what the theory means in any intuitive or common sense way. 'Interpretations of Quantum Mechanics' seek to provide a way of picturing what is going on in a quantum system.

Any mathematical theory needs an interpretation which allows scientists and the public at large to have a mental image of what it means. But the interpretation is distinct from the mathematical equations which are the essence of the theory: the interpretation may be modified or completely changed while retaining the mathematical form of the theory. The interpretation of **Maxwell's theory of electromagnetism** originally involved a luminiferous **aether** — a substance pervading all space which supported **electromagnetic wave** motion (§1.5). In the 19th century a wave was understood as the collective motion of particles in a medium such as air for sound waves or water for water waves, so it was natural to expect there to be a medium through which electromagnetic waves propagate. Maxwell himself attempted to construct mechanical models of the aether, but it became clear that it would have to have self-contradictory properties: it had to behave like a solid in order to transmit transverse waves and at the same time act as a transparent fluid pervading all space yet not impeding the motions of the planets! However, the theory embodied in **Maxwell's equations** retained its significance after the concept of the luminiferous aether was abandoned.

The mathematical model that is Quantum Theory is similar to Maxwell's equations in opening up new areas of understanding and experimentation, while interpretations of the theory are like the ideas of the luminiferous aether — full of apparent contradictions and difficulties, though useful for providing a mental picture of what is going on. The mathematical structure may be used to explain and predict phenomena without recourse to any detailed or coherent interpretation, just as electromagnetic phenomena such as the reflection and refraction of light at a boundary between different media may be derived from Maxwell's equations without any reference to the existence or properties of an aether.

Not all scientists using Quantum Theory subscribe to exactly the same interpretation: many cosmologists prefer the idea of the **multiverse** (arising from the **Everett many-worlds interpretation**), other physicists retain a version of the (rather nebulous) **Copenhagen Interpretation,** and there are some who search for new interpretations. An interpretation is a philosophical aid to those using or attempting to explain the theory, and it remains just a philosophy rather than an essential part of the theory unless it becomes susceptible to experimental test. The Michelson–Morley experiment put paid to the 19th century concept of the aether, and the **Special Theory of Relativity** showed that the null result of that experiment could be explained through the postulate of the velocity of light as a limiting velocity. More recently **Quantum Field Theory** has provided a picture of light in terms of **photons,** and electromagnetic fields as clouds of virtual photons. The luminiferous **aether** is replaced by the quantum vacuum which seethes with virtual particles continually being created and destroyed. So a combination of experiment and new theories combined to transform the interpretation of electromagnetism. But quantum waves are no less hard to picture than were Maxwell's waves for the 19th century physicists. Maxwell's were at least waves in physical space, whereas quantum waves are waves in multi-dimensional **configuration space**. (There are 3N spatial components in the configuration space of N particles.) It might seem that quantum waves are physically real because they produce interference effects, yet in relating them to physical observations they must be understood as probability waves (§7.3.9).

Born's view of the wave function as a **probability amplitude** led him to suggest an **ensemble interpretation**: Quantum Mechanics applies only to ensembles of particles, and the wave function determines the statistical behaviour of the ensemble. Certainly it is not possible to compare the predicted probability distribution with experiment without repeating the same experimental procedure on a large ensemble of identically prepared particles. However, in agreement with Bohr's ideas, the behaviour of an individual particle cannot be pictured. In the ensemble interpretation the wave function does not determine the behaviour of single particles. In this type of interpretation there is a fundamental **indeterminacy** in nature. Einstein believed that this meant that Quantum Mechanics is incomplete, and that there is a **deterministic theory** underlying it. A deterministic approach was originally suggested by **de Broglie**, but this was eclipsed by Bohr's Copenhagen Interpretation. The argument between those who believe in a fundamentally statistical interpretation and those who look for a deterministic theory underlying Quantum Mechanics continues to this day.

12.1.1. *The measurement problem revisited*

The measurement problem, §10.4, lies at the centre of the difficulties in finding an acceptable interpretation of Quantum Mechanics. According to Bohr's Copenhagen Interpretation, measurements on quantum systems inevitably involve macroscopic events, such as the movement of a pointer, which behave according to classical physics. When a measurement is performed quantum uncertainty concerning the outcome of the measurement is resolved — the **wave function collapses** (§7.3.9). There is a boundary between the quantum and the classical realm, with **indeterminacy** and **entanglement** on one side and and the classical behaviour of our every day experience on the other. But, as Bohr insisted, this boundary is not fixed. It depends on the observer's choice of which (classical) measurement to perform.

Schrödinger's cat (§10.4.1) highlights this difficulty: if the classical measurement is made when the observer opens the box, then the cat will have been in an indeterminate state before this. **Wigner** made the point even more forcefully (Wigner, 1967) by having a friend of the observer — **'Wigner's friend'** — present at the cat experiment. The observer leaves the room while the cat is still shut in the box, and while he is away the friend opens the box and sees the cat to be either alive or dead. However, the original observer does not know what his friend has seen. To him the cat is still in a superposition of live and dead states until he returns and discovers what his friend has observed. Wigner suggested that this implies that human consciousness is what causes the wave function to collapse: Wigner's friend causes the indeterminate state of the cat to collapse. The absent observer is uncertain about the cat's state, but this uncertainty no longer reflects a real indeterminacy in the state of the cat once Wigner's friend has opened the box. And what if the cat were replaced by Wigner's friend? If human consciousness controls the collapse of the wave function he will always be in a classically well-determined state, though the observer outside the box will not know whether he is alive or dead. Does the boundary between classical and quantum shift according to whether or not a human observer is present? Before the advent of humans, did wave functions associated with events in the universe never collapse?

The problem is solved to some extent by **decoherence**. When a measurement is made on a quantum particle it interacts with particles in the detector and becomes **entangled** with them. During this measurement the environment is not monitored by the observer and information about the interference terms in the quantum probability distribution, which are due to **coherence**, is lost. This results in the probability distribution for the results of the measurement being reduced to a classical one. More significantly, the environment itself can lead to decoherence without the intervention of a human observer. Any quantum particle becomes entangled with very large numbers of particles through interactions with its environment, and the coherence associated with the particle effectively 'leaks out' into the environment. Models of the environment as a scalar field with thermal fluctuations suggest that interactions with it can lead to the loss of interference between states of the particle corresponding to macroscopically different positions. In this case the probability distribution for the position of the particle reduces to a peak at each possible position, with height identifying the probability of finding the particle there (Zurek, 2002). This means that decoherence can produce states that look classical, in the sense that a particle is predicted to be found at one of a set of macroscopically distinct positions. However, it is not possible to predict which of the possible positions it will be found at. There is no mechanism which explains how an individual measurement selects a particular result out of the set of all possible results. The pseudo-classical probability distribution hides a fundamental indeterminacy in the outcome of any measurement. The question remains whether there is a deterministic theory underlying Quantum Mechanics, as Einstein believed, or whether 'God plays dice'.

12.1.2. *The significance of Bell's work*

John Bell's work, discussed in Chap. 10, led to a renewed discussion about the interpretation of Quantum Theory. In the 1960s, when Bell started his seminal work, it was heresy to question the **Copenhagen Interpretation**. Students were taught that they must not attempt to understand what is happening at a quantum level: we can only know the results of classical measurements on quantum systems; Quantum Mechanics allows us to predict the results of these measurements, but our classical minds cannot possibly picture the behaviour of quantum particles, so we should not try to do so. For instance, it is meaningless to talk about the trajectory of an electron or photon in a **two-slit experiment**. Indeed, originally, Bohr had said that his principle of **complementarity** means that the wave and particle aspects of a quantum particle cannot both be observed in the same experiment. With the advent of single particle detectors it became clear that in a two-slit experiment both aspects are apparent: the arrival of individual particles can be detected at the final screen, but at the slit screen the wave aspect dominates, and determines the probability distribution of arrival points at the detector.

There were attempts to probe the behaviour of the quantum world, notably by **Bohm** in the 1950s, who resurrected and developed de Broglie's **pilot wave theory** (discussed in Chap. 11 and §12.4). Bell was inspired by this theory to investigate the possibility of **hidden variables** (Bell, 2003: Paper 17). He focussed on the **Einstein–Podolsky–Rosen** discussion (§9.4.2 and §10.1) and the question of **locality**. His inequality, in various formulations, can be violated by the predictions of Quantum Mechanics, indicating that Quantum Theory is non-local. More importantly the inequality can be tested experimentally. Many experiments on entangled pairs of particles have shown that nature violates the inequality and behaves as predicted by Quantum Mechanics. Bell's work and its experimental tests spawned huge research interest associated with entanglement and the modern field of quantum information theory described in §10.3. So probing the implications of an interpretation can lead to great advances if experimental tests of those implications can be devised. Testing an interpretation against nature brings it into the realms of physics rather than philosophy.

Current work continues to probe the implications of Quantum Mechanics and search for refined or new interpretations. The combination of theory and experiment is essential to further advance our understanding. Some aspects of current theoretical work are discussed in this chapter.

12.2. The Multiverse

The measurement problem poses a particular difficulty for cosmologists, since it seems that the whole universe should be described by a quantum mechanical wave function. This wave function will inevitably contain terms referring to mutually exclusive outcomes to observations. By definition, if we are talking about the universe itself, there can be no external observer as required by the Copenhagen interpretation, and the question also arises as to what constituted an observation in eras before the appearance of sentient observers. Furthermore, astronomical and cosmological observations are not experiments that can be repeated under identical conditions in order to measure the probability distribution of results. If there is a single wave function describing the state of the universe, how is it that events in the universe appear to be unique?

The modern 'multiverse' theory stems from **Everett's many-worlds interpretation** (§10.4.1). If a wave function always evolves according to the **Schrödinger time-dependent equation** then it cannot collapse, and the only alternative is that all possible outcomes of a measurement co-exist in different worlds. The totality of different 'worlds' is the **multiverse**. Its wave function never collapses, but all possible outcomes of any measurement or observation exist in different universes within the multiverse. Our universe is just one branch of the multiverse. Any observer is only aware of the branch he inhabits, though he will have alter egos in other branches who observe slightly different universes. Other universes within the multiverse might be regions of space beyond our cosmological horizon (which marks the limit of the observable universe), they might be regions with different values of fundamental constants, or even with different laws of physics. More abstractly, they might be different branches of the wave function of the multiverse, which exist in the multidimensional **Hilbert Space** of quantum states. **Decoherence** may be invoked to explain which aspects of the universe appear to behave classically (Tegmark, 2010).

Cosmological models can describe a variety of universes with different characteristics, many of which would not sustain life as we know it. This leads to the 'anthropic principle': it seems that we inhabit a universe that is fine-tuned for the existence of life. If our universe were the only one it is strange that it should possess exactly the properties necessary for us to exist. On the other hand, in a multiverse there could be many universes inhospitable to life as well as others in which life can develop. From this point of view it is not surprising that we find ourselves in one of the latter type. **String theory**, which is a major candidate for combining gravitation and Quantum Theory, has a huge number of different solutions, with a large variety of physical parameters. The current understanding of string theory does not permit one solution to be preferred over others, and so the different solutions might be seen as a range of different universes co-existing within the multiverse (Weinberg, 2005).

It is impossible to verify the multiverse theory through direct observation of other universes, since there is no interaction between different universes within the multiverse, and we can never observe any universe apart from our own. However, Tegmark believes that

> "Any multiverse theories can be tested and falsified, but only if they predict what the ensemble of parallel universes is and specify a probability distribution over it."
>
> (Tegmark, 2010)

Cosmological and string theories describe a range of universes with different values for fundamental quantities such as the curvature of space or the energy of the vacuum. They can identify the probability of finding the values we observe in our own universe. The anthropic principle restricts the values we could possibly observe to those consistent with a universe in which life can develop. So observations of values consistent both with the range predicted by these theories and with the anthropic principle might be understood to support the idea of the multiverse.

A different approach to the multiverse is taken by **David Deutsch**. He believes that it can be simulated by a network of **quantum computers**. What matters is information rather than particles or waves, and information flow can be modelled as **qubits** travelling through a network of quantum gates. He claims that there is no non-locality problem with information. For example, in a Bell-type experiment with entangled pairs of particles (such as that described in §10.2.2) the spatially separated observers at D_A and D_B both see a random series of hits and misses. The information about the correlation between results at detectors D_A and D_B is not available to the observer at D_B until the observer at D_A sends a record of the settings he has used. This information is sent through classical channels. So the identification of the correlations depends on the information observed at detector D_B and the information sent from the observer at D_A. There is no instantaneous non-local influence on the *information* obtained from the results at D_B. A similar argument was used in §10.3 to show that Special Relativity is not violated in such experiments, since information cannot be exchanged at a speed greater than that of light. As long as we think in terms of information rather than particles, non-locality is not a problem (Deutsch, 2002).

12.3. A Deterministic Local Field Theory

It is generally accepted that the violation of Bell's Inequality means that there can be no local hidden variables theory underlying Quantum Mechanics. However 't Hooft is working on a deterministic local field theory from which Quantum Mechanics would arise as a statistical theory for events at the much larger scale of quantum particles, which we shall refer to as the 'atomic' scale ('t Hooft *et al.*, 2005).

Standard **Quantum Field Theory** is already local in the sense that changes in the field at a particular point in space-time are determined by the behaviour of the fields infinitesimally close to that point. The fields are continuous functions of space and time, and changes in them cannot propagate faster than the speed of light. However, the theory is not deterministic. General Relativity indicates that at the **Planck scale** (distances of the order of 10^{-35} m and times of the order of 10^{-44} s, Appendix J) a new understanding of space and time is needed, and space-time may even be discrete rather than continuous. So at distance scales at and below the Planck scale Quantum Field Theory is likely to need modification, and a theory of **quantum gravity** is essential. 't Hooft believes that the difficulties in finding a completely satisfactory theory of quantum gravity are fundamental, and cannot be solved without a new approach to Quantum Theory. So he is examining an underlying deterministic quantum field theory at some very small scale such as the Planck scale, for which Quantum Mechanics is the appropriate statistical theory at the very much larger atomic scale. He has investigated model theories on a **discrete space-time**. His local deterministic theory would not be a hidden variables theory in the usual sense since it is a field theory. The fundamental 'hidden variables' would be the degrees of freedom of the deterministic field rather than hidden characteristics of quantum particles.

't Hooft compares Quantum Mechanics to **thermodynamics**. Thermodynamics gives a statistical description of gases on a scale much larger than that of the constituent molecules. It is not useful to attempt to trace the trajectories of individual particles even if their behaviour is assumed to be governed by local and deterministic classical laws, and thermodynamic variables such as temperature and entropy are not well defined at the much smaller molecular scale. In a similar way quantum particles would obey the statistical laws of Quantum Mechanics even if there is an underlying deterministic local field theory at the Planck scale. From this point of view, a particle such as an electron or photon and its position, momentum, energy and spin are no more 'real' than in the **Copenhagen Interpretation**. They are defined only by the statistical theory, just as temperature and entropy are defined by thermodynamics. However, if there is a local deterministic theory underlying Quantum Mechanics, the outcome of any measurement on a quantum particle is not indeterminate. Branches of the quantum probability distribution representing results which are not observed merely represent our lack of knowledge, in contrast to the many worlds interpretation. 't Hooft's attitude to Quantum Mechanics is that it is

> "the theory enabling us to produce the best possible predictions for the future, given as much information as we can give about the system's past, in any conceivable experimental setup. Quantum mechanics is not a description of the actual course of events between past and future."
>
> ('t Hooft, 2007)

In 't Hooft's underlying deterministic local field theory all degrees of freedom are classical in that they can have simultaneous well-defined values, and the fields evolve according to deterministic dynamical laws. However, deterministic dynamical laws can produce chaotic states, and 't Hooft suggests that the vacuum might be a chaotic dynamical state. In this case the behaviour of quantum particles at the atomic scale would be reminiscent of **Brownian motion**: pollen grains suspended in liquid appear to move in a random manner because they are continually buffeted by the much smaller molecules of the liquid. Even if we assume that the motion of the molecules is deterministic, dynamical behaviour on the much larger scale of the pollen grains can only be treated statistically. Similarly, the behaviour of quantum particles would have to be treated statistically even if the vacuum surrounding them is actually a deterministic, though chaotic, state.

It is necessary to define a **Hamiltonian operator** to control the way a state evolves in time (Eqn. 7.69). One difficulty 't Hooft has encountered is that in general the eigenvalue spectrum of a Hamiltonian derived from an underlying deterministic theory does not have a lower bound. That is, there would be no lowest energy, or ground state, and no way of defining a vacuum state. A quantum theory without a vacuum state would be as unrealistic as thermodynamics without an absolute zero of temperature. This difficulty can be overcome if there is some information loss in moving from the deterministic to the statistical description. This could be due to many different states at the primordial Planck level evolving (deterministically) into states which cannot be distinguished from one another. So the states in the statistical theory would have to be groups ('equivalence classes') of primordial states rather than the primordial states themselves. The information which distinguishes the states in the same equivalence class is lost ('t Hooft, 2007).

If space and time are quantised rather than continuous at the Planck level it is necessary to account for **symmetry laws,** such as translational, rotational and Lorentz invariance, which are defined in terms of continuous space-time transformations. An operator representing a discrete translation in space may be extended mathematically to represent fractional translations even if such translations are not physically realisable at the Planck scale. Similarly continuous translations may be defined mathematically by taking the limit in which the transformations are infinitesimal. According to 't Hooft, the fractional transformations would transform states on a discrete space-time into superpositions of such states. So the basis states of the statistical field theory at the atomic scale, in which space and time are treated as continuous, might be entangled states. The symmetry laws would not be symmetries of the underlying theory, but would emerge as symmetries of the statistical theory ('t Hooft, 2007). The idea that space-time symmetries are emergent rather than fundamental has also been proposed recently as the basis for a new approach to a theory of quantum gravity (Hořava, 2009).

Although 't Hooft's underlying theory would be local, he envisages that the statistical theory at the atomic level would be identical to Quantum Mechanics. Thus, measurements on entangled quantum particles would violate Bell's Inequality in the way predicted by standard Quantum Mechanics. Locality of an underlying field theory does not imply that the theory governing behaviour at the level of quantum particles is also local.

12.4. A New Look at the de Broglie–Bohm Pilot Wave Theory

Louis de Broglie was awarded the 1929 Nobel Prize in physics "for his discovery of the wave nature of electrons". The development of his ideas is discussed in detail in Chap. 11. In his doctoral thesis in 1924 (§11.3) he had suggested that not only light but also particles such as electrons should show both a wave and a particle aspect, and predicted the diffraction of electrons. He had also explained the quantisation condition for orbital electrons in terms of an associated standing wave (§7.2). By 1927 his ideas had progressed, and he described his 'pilot wave' theory at the **Solvay Conference** that year (§11.7 and §11.9). He emphasised a full dualism between the particle and wave aspects: the particle was localised within the extended region occupied by the associated pilot wave. The trajectory of the particle was determined by a dynamical equation which incorporated the effect of the pilot wave, and the wave satisfied Schrödinger's equation. However, Schrödinger preferred to understand the wave as the fundamental reality, and hoped to explain the particle as a wave packet. De Broglie's view of particle–wave duality was in complete contrast to Bohr's, since Bohr rejected the reality of particle trajectories. By the end of the 1920s Bohr's interpretation was triumphant, and de Broglie capitulated to the majority view.

In the 1950s Bohm resurrected de Broglie's pilot wave theory in a new form (Bohm, 1952; Bohm and Hiley, 1993), and later Bell referred to the theory in published papers and believed that it should be more widely known (Bell, 2003). Although the theory is deterministic and can identify individual particle trajectories it is non-local, and its predictions violate Bell's Inequality in exactly the same way as the predictions of standard Quantum Mechanics. More recently Valentini has returned to the de Broglie–Bohm theory and treated it as a credible candidate for a deterministic non-local hidden variables theory. An important feature of his work is that he has suggested observations which might test it (Valentini, 2009).

12.4.1. *The essential ideas of de Broglie's pilot wave theory*

Valentini has re-examined de Broglie's work and shown that it has been significantly underestimated and ignored (Bacciagaluppi and Valentini, 2009). In giving equal status to the particle and wave aspects de Broglie sought a **new dynamics** which would combine **Fermat's Principle** of Least Time with **Maupertuis' Principle** of Least Action (§7.3.4, §7.3.5 and §11.3). In optics, Fermat's extended principle, Eqn. 7.25, relates the path of a light ray to the wavelength λ and frequency v of the light wave. At each instant the direction of the ray is determined by the gradient of the phase of the light wave, so that the ray is always perpendicular to the wavefront. (The phase of a light wave at position r and time t is $2\pi(v'r - v t)$, where $v' = 1/\lambda$ is the wavenumber, and a wavefront is a curve on which this phase takes a constant value.) The relationships between the mechanical and wave-like characteristics of a particle are given by Einstein's expression $E = hv$, Eqn. 7.11, and de Broglie's $p = h/\lambda$, Eqn. 7.16. Inserting these in Eqn. 7.25, Fermat's Principle is transformed to Maupertuis', Eqn. 7.30a. So the waves associated with a free particle with momentum p and energy E have a phase $2\pi(p.r - Et)/h$. The direction of the momentum corresponds to the direction of the ray in optics, so that it is natural to identify the momentum with the gradient of the matter wave.

In terms of his pilot wave theory, reported at the 5th Solvay Conference, de Broglie viewed particles as being 'guided' by pilot waves (§11.9). In accordance with the relationship between Fermat's and Maupertuis' principles, the momentum **p** of a particle is determined by the gradient of the phase of its associated pilot wave:

$$\mathbf{p} = m\mathbf{v} = \nabla S. \tag{12.1}$$

Here, m is the mass of a particle with velocity $\mathbf{v}$, and S is (apart from a constant factor $\hbar = h/2\pi$) the phase of the pilot wave $\Psi = |\Psi| \exp(iS/\hbar)$. The wave satisfies Schrödinger's equation, Eqn. 7.69. (In fact Schrödinger discovered his famous equation as a result of his search for a wave equation to describe de Broglie's waves, §7.3 and §11.3). Equation 12.1 is a '**guidance condition**' tying the particle trajectory to the phase function of the wave.

In his 1927 theory de Broglie extended the one particle model to cover a system of particles. Particles and waves have a distinct physical existence, but their dynamical evolution is coupled through equations like Eqn. 12.1 for the velocity of each particle and the Schrödinger equation for the pilot wave. Although the particles move in physical space the wave function is a wave in the multi-dimensional configuration space of all the particles in the system. Each particle has a hidden but well-defined trajectory. This is in contrast to Bohr's belief that trajectories do not exist, and that since they are indeterminate one should not attempt to picture them.

Bohm differentiated de Broglie's Eqn. 12.1 for the velocity to obtain an equation for the acceleration $\mathbf{a}$, and thus for the force on the particle. This equation can be written

$$m\mathbf{a} = -\nabla (V + Q), \tag{12.2}$$

where V is the usual potential which appears in the Schrödinger equation and Q is a new '**quantum potential**' which is determined by the second derivative of the modulus of the wave function:

$$Q = -\frac{\hbar^2}{2m} \frac{\nabla^2 |\psi|}{|\psi|}. \tag{12.3}$$

If the initial position of an individual particle is specified, its trajectory may be determined through Eqn. 12.2. The effect of the quantum potential is non-local since it depends on the pilot wave which is

Figure 12.1. Trajectories of particles in a two-slit experiment.
(Reproduced from Bohm and Hiley (1993) with permission of the publisher).

spread over an extended region of space. This non-locality produces strange kinks in the trajectories, as can be seen in Figure 12.1 which shows the predicted trajectories of particles in a **two-slit diffraction experiment** (Bohm and Hiley, 1993: Figure 3.1).

Bohm treated de Broglie's Eqn. 12.1 as a constraint on the initial momenta which could be relaxed. His Eqn. 12.2 frames the theory in standard Newtonian form. On the other hand de Broglie set out to create a new dynamics based on uniting Fermat's and Maupertuis' principles, and this novel idea is lost in the Bohm version. Of course de Broglie's dynamics as well as Bohm's is **non-local**, and thus a suitable candidate for a **hidden variables theory** which violates **Bell's Inequality**.

In classical dynamics a knowledge of the initial conditions is needed to derive a unique trajectory for a particle. Similarly, in de Broglie's and Bohm's theories the trajectories of the particles depend on their initial dynamical states, which are the hidden variables of the theory. The initial state of both particles and waves must be specified. Normally, a quantum measurement is performed on an ensemble of particles which have all been produced in a similar way but which have different initial positions, such as electrons in a beam with well-defined energy and momentum used in an electron diffraction experiment. The initial state of the ensemble of particles is given by their initial distribution in space, $\rho(0)$, and that of the pilot wave by the initial value of the wave function, $\Psi(0)$. De Broglie coupled these two initial states, identifying the initial distribution of the particles with $|\Psi(0)|^2$ (§11.7 and §11.9). If this is the **probability distribution** of the particles at one instant, the Schrödinger equation, which determines the time evolution of the wave function, guarantees that $|\Psi(t)|^2$ will be the probability distribution at any other time t. So de Broglie's initial condition means that the pilot wave satisfies the **Born interpretation** of the wave function (§7.3.9), and the pilot wave model leads to what is now standard Quantum Mechanics. However, the **measurement problem** is avoided: the wave function does not collapse, but any particular particle exists in only one of the possible branches. The different possible results are just that — possibilities for the results of measurements on an ensemble of particles whose initial probability distribution has been specified by the wave function. The initial condition for an individual particle would determine the outcome of the measurement on it uniquely, but experimentally the precise position of a particle cannot be determined — it is a hidden variable.

12.4.2. *Valentini's development of the de Broglie–Bohm theory*

In his re-examination of the de Broglie–Bohm theory, Valentini focusses on the choice of initial distribution, $\rho(0)$, for the particles. He points out that it need not necessarily be the same as the Born distribution $|\Psi(0)|^2$, though it would be expected to relax to the Born distribution as the system evolves in time. In **thermodynamics** a non-equilibrium thermal distribution of particles relaxes to thermal equilibrium, and Valentini compares standard Quantum Mechanics to equilibrium thermodynamics, and identifies the **Born distribution** as an **equilibrium quantum distribution**. Identifying the initial distribution of the particles with the initial value of the modulus squared of the wave function is similar to restricting oneself to states which have already reached thermal equilibrium. The important field of non-equilibrium thermodynamics deals with ensembles which are not in thermal equilibrium, and in which it is possible to observe transfer of thermodynamic variables, such as temperature, between different parts of the system. Valentini suggests that the existence of a **deterministic hidden variables theory** underlying Quantum Mechanics could be revealed by experimental evidence for particles with an initial probability distribution *different* from the Born distribution (Valentini, 2009).

In a non-equilibrium quantum distribution the non-locality of the de Broglie–Bohm model means that faster-than-light interactions between particles could be observed. In quantum equilibrium these are masked by 'quantum noise' just as heat exchange between individual parts of ensembles in thermal equilibrium is masked by thermal noise. If one could produce particles with a non-equilibrium quantum distribution it would be possible to exchange information instantaneously. In §10.3 we argued that the random nature of the results of a polarisation measurement on one photon of an entangled pair prevents the instantaneous transmission of information to an observer making a polarisation measurement on the remote photon of the pair. But if one could use entangled pairs in a suitable non-equilibrium state the statistics of the results on the first photon could be different from $50:50$ and information might be transmitted to the second observer instantaneously.

But how can a source of non-equilibrium quantum states be obtained? To produce such states in the laboratory is probably impossible, but Valentini suggests several astrophysical and cosmological tests (Valentini, 2007). In the early universe, interactions between particles would have led to quantum equilibrium, but perhaps the rapid expansion of the universe in the inflationary period might have increased the mean free path of some particles sufficiently rapidly that they did not have time to relax to the equilibrium state. Evidence for such relic non-equilibrium states might be found in the temperature fluctuations of the cosmic microwave background. Alternatively, Valentini has suggested that gravitational effects might be able to generate quantum non-equilibrium states today. He proposes looking for evidence for this in Hawking radiation from black holes and in neutrino oscillations. Finally, he suggests that quantum probabilities should be measured at the smallest possible length scales, since gravitational effects at the **Planck Scale** (Appendix J) could possibly generate quantum non-equilibrium.

12.5. Where Next?

Interpretations of Quantum Mechanics vary widely, but their basic assumptions fall into one of two main groups: on one hand are those which assume that Quantum Theory is complete and fundamental, and on the other those which assume the existence of a theory underlying Quantum Mechanics. The **Copenhagen** and **multiverse** interpretations belong to the first group, while **deterministic hidden variables** interpretations belong to the second. Any assumption that there is no underlying theory implies that the statistical nature of Quantum Mechanics reflects a fundamental **indeterminacy** in nature.

The Copenhagen Interpretation, though it is not a coherent picture, allowed Bohr and others to develop rules for extracting predictions from the mathematical theory. These rules, such as **Born's probability interpretation** of the wave function and **wave function collapse**, remain essential for using Quantum Mechanics to make predictions. The multiverse interpretation takes the assumption that Quantum

Theory is complete to the extreme limit in which the wave function cannot evolve except as prescribed by the Schrödinger equation. This means that the collapse of the wave function cannot occur. It only appears to occur because we observe the multiverse from within one of its component universes.

Any deterministic interpretation assumes a theory underlying Quantum Mechanics. A field theory might be both local and deterministic, as suggested by 't Hooft. In spite of the fact that his theory would imply a unique outcome for any measurement, it does not provide an intuitive picture of quantum behaviour at the 'atomic' scale of electrons, photons and other quantum particles. Such particles are artifacts of the statistical theory. Their behaviour would be completely described by Quantum Mechanics, and consequently Bell's Inequality would be violated in spite of the locality of the underlying deterministic field. Experimentally observed violations of Bell's Inequality imply that a theory that is deterministic at the level of quantum particles cannot be local, and it seems likely that no satisfactory interpretation at this scale can be both deterministic and local.

A theory such as that of de Broglie and Bohm introduces an important assumption, that the particle aspect should be treated on an equal footing with the wave aspect. In standard Quantum Mechanics the wave aspect is pre-eminent: particle dynamics appears only through the Hamiltonian, which determines the evolution of the wave function. In de Broglie's formulation the particle and the pilot wave each obey their own equation of motion: the particle has a trajectory determined by Eqn. 12.1 and the wave evolves according to Schrödinger's Eqn. 7.69. The behaviour of particle and wave are coupled since the Schrödinger equation includes parameters associated with the particle (in the Hamiltonian) and Eqn. 12.1 relates the momentum of the particle to the phase of the wave. Although each particle trajectory is unique even if 'hidden', it would look very different from a classical one, and depend on non-local effects. The question remains as to the nature of these non-local effects — what sort of interaction does Bohm's quantum potential, Eqn. 12.3, represent?

It is very difficult to find ways of testing any of the interpretations discussed here. According to Tegmark:

> "Multiverses are not theories but predictions of certain theories, and such theories are falsifiable as long as they also predict something that we can test here in our own universe."

(Tegmark, 2010)

The theories he refers to are **cosmological theories** and **string theories**. He suggests that cosmological tests of the degree to which our universe is fine-tuned for the existence of life can support the multiverse interpretation. However, these tests cannot involve physical observations of other universes within the multiverse.

Valentini also suggests cosmological tests. In his case these rest on a specific assumption, that the distribution of particles is not necessarily identical to the **Born distribution**. If incontrovertible evidence for quantum non-equilibrium states were found it would provide a very precise indication that current understanding of the wave function needs to be amended, and that a hidden variables theory might underly Quantum Mechanics.

The difficulties in providing a universally acceptable interpretation of Quantum Mechanics suggest that we have not yet reached a satisfactory understanding of it. Is there any chance that a completely new point of view could throw fresh light on the problem? If so, perhaps it will come from a deeper understanding of space and time. The **General Theory of Relativity** revolutionised our view of space-time, describing it as a dynamical structure with local curvature depending on the matter and radiation in that region. At distances small compared to the **Planck scale** the curvature can increase indefinitely, and the very existence of space-time becomes unclear, leading to the possibility that space-time may be quantised at this scale. **Symmetry principles** which have been taken as fundamental at the smallest possible scales might instead emerge from the theory only at larger scales (§12.3). In theories that seek to combine gravitation and Quantum Theory, such as string theory, the concept of space-time should emerge from the theory. But at a less esoteric level it is hard to break the habit of thinking of it as a pre-existing framework, and indeed it is treated as such in standard Quantum Mechanics.

Problems of **determinism** and **non-locality** become evident on the scale of quantum particles such as electrons and photons whenever spatial characteristics are measured. For example, in a **two-slit experiment** the wave associated with the particle passes through both slits, but the particle cannot split and pass through both. So the position of the particle at the slit screen is indeterminate — any attempt to define it sufficiently to identify which slit the particle passed through destroys the interference pattern. In a measurement of the spin of an electron or polarisation of a photon it is the spatial orientation of an internal characteristic of the particle that is being measured. In contrast, internal characteristics, such as the mass, electric charge and total spin of a free electron, are well defined. The spin is exactly 1/2, and the accuracy with which the mass and charge can be measured is limited by experimental factors, not by a fundamental indeterminacy in their values. Could it be that the indeterminacy in our measurements of the spatial characteristics of free particles is due to an indeterminacy of space itself, rather than of the particle?

To answer that question requires an understanding of how space and time are identified experimentally. How do we define a coordinate system within which we can identify position and direction? Macroscopically space defines the relationships between a variety of different physical objects which we observe through the light emitted from or reflected by them. We use it as a labelling system. But to maintain the spatial relationship between two objects they must interact with each other and with other objects in their environment. It might be that the very structure of space and time is not determined at a fundamental level. Perhaps the concept of a well-defined space-time emerges only at a level where a sufficient number of particles are interacting with one another. If a particle is not interacting with others, at least gravitationally, its position and orientation in space cannot be defined. If space were a macroscopic labelling system, determined by the mutual interactions of a large ensemble of material objects and particles, then it would not be surprising if a single particle (such as a photon or electron) cannot be uniquely labelled until it has interacted appropriately with the ensemble which defines space. A classical measurement of an appropriate kind would provide the necessary interaction. From this point of view it would be the identification of a coordinate system that is determined by a statistical theory as opposed to the characteristics of a particle.

These are merely vague speculations. Perhaps a clearer understanding of the relationship between gravitation and Quantum Theory will provide a completely new interpretation of Quantum Mechanics.

12.6. Further Reading

Bacciagaluppi, G. and Valentini, A. 2009.
  Quantum Theory at the Crossroads: Reconsidering the 1927 Solvay Conference, Cambridge: Cambridge University Press.
Bell, J.S. 2003.
  Speakable and Unspeakable in Quantum Mechanics, 2nd edn. Cambridge: Cambridge University Press.
Bohm, D. 1952.
  A Suggested Interpretation of the Quantum in Terms of 'Hidden' Variables. I. *Phys. Rev* 85: 166–179.
Bohm, D. and Hiley, B.J. 1993.
  The Undivided Universe. London: Routledge.
Deutsch, D. 2002.
  The Structure of the Multiverse. *Proc. Roy. Soc. Lond.* A **458**: 2911–2923.
Hořava, P. 2009.
  Quantum gravity at a Lifshitz point. *Phys. Rev. D* **79** 084008. (Also reported by Ananthaswamy A. 2010. Rethinking Einstein: the end of space-time. *New Scientist* 207(2772): 28–31).
Tegmark, M. 2010.
  In S. Saunders, J. Barrett, A. Kent and D. Wallace (eds), *Many Worlds? Everett, Quantum Theory and Reality*. Oxford: Oxford University Press, and arXiv:0905.2182v2.
't Hooft, G., Witten, E., Dowker, F., and Davies, P. 2005.
  Does God Play Dice? *Physics World* 18(12): 21–23.
't Hooft, G. 2007.
  Emergent Quantum Mechanics and Emergent Symmetries, *AIP Conference Proceedings* 957: 154. Also presented at the 13th International

Symposium on Particles, Strings and Cosmology, PASCOS, Imperial College London, July 6, 2007, arXiv:0707.4568v1.

Valentini, A. 2007.
Astrophysical and cosmological tests of quantum theory. *J. Phys. A: Math. Theor.* **40**: 3285–3303.

Valentini, A. 2009.
Beyond the Quantum. *Physics World* **22**(11): 32–37.

Weinberg, S. 2005.
Opening talk at the symposium Expectations of a Final Theory, at Trinity College, Cambridge, on 2 September 2005. Published in Bernard, C., ed. (2007) *Universe or Multiverse?* Cambridge: Cambridge University Press.

Wigner, E. 1967.
Symmetries and Reflections. Bloomington: Indiana University Press.

Zurek, W. 2002.
Decoherence and the transition from quantum to classical – Revisited. *Los Alamos Science* **27**. Available at http://arXiv.org/pdf/quant-ph/0306072. This is an updated version of *Physics Today* (1991) **44**: 36–44.

Appendix A

Entropy

"Maximum disorder was our equilibrium"
(T. E. Lawrence in *The Seven Pillars of Wisdom* —
quoted by H. A. Bent in *The Second Law*)

A.1. The concept

The concept of entropy (introduced by Clausius in 1865) can be readily appreciated by enquiring into what characterises the adiabatic in the **Carnot Cycle**. We therefore need to recall that in the ideally perfect simple heat engine [Fig. A.1(a)], which was the concept of Sadi Carnot in 1824, a 'working substance' (its nature is irrelevant) that can be compressed or expanded by a piston P is taken through a cycle of changes starting in contact with a cold 'sink' at temperature T_2 [point A in Fig. A.1(b)]:

  (i) From the cold sink it is placed on an insulator and compressed adiabatically (A $\rightarrow$ B) so that its temperature rises (corresponding to the amount of work done on it) till it is at the temperature T_1 of a hot body.

 (ii) It is then moved to the hot body and allowed to expand (B $\rightarrow$ C) isothermally, absorbing from the hot body the amount of heat Q_1 required to do so.

(iii) After being returned to the insulator, it is allowed to expand adiabatically till its temperature drops back to T_2 (C $\rightarrow$ D).

(iv) Finally, back on the cold sink, it releases an amount of heat Q_2 in an isothermal compression to take it from D $\rightarrow$ A, its starting point.

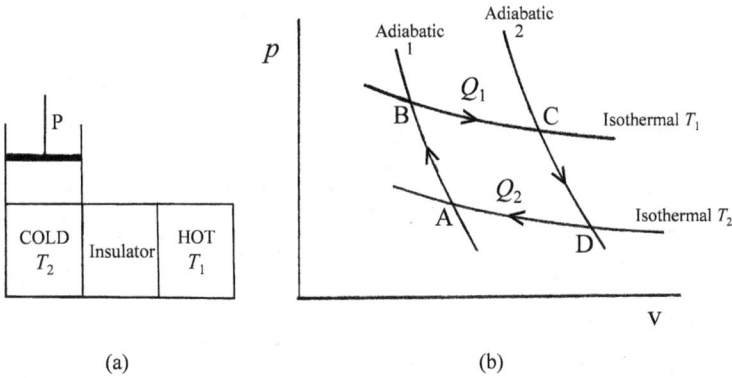

Figure A.1. Carnot cycle.

The mechanical work done, W, in the cycle represented by the area ABCD is equal to $Q_1 - Q_2$. The efficiency of the engine, per cycle, is defined as

$$\text{Efficiency} = \frac{\text{external work done}}{\text{heat taken from the hot body}}$$

$$= \frac{Q_1 - Q_2}{Q_1}. \tag{A.1}$$

Thus, the heat engine works by extracting heat Q_1 from a hot body, some is converted into work, and the remainder, Q_2, must be given up to a colder body. The greater the difference between Q_1 and Q_2, the greater the amount of the available energy for work.

Likewise, it follows that it is impossible to cause heat to pass from one body to another at a higher temperature without the aid of a supply of energy. This is Clausius' statement of the **Second Law of Thermodynamics**. It corroborates the notion which cannot be proved experimentally, that heat cannot flow spontaneously from a cold body to a hotter body.

We need to note the following:

(i) The efficiency of a Carnot cycle operating between two temperatures is independent of the quantity of heat absorbed from the hot source. This is shown by doubling the area of the cycle in Fig. A.1(b) which simply gives the efficiency $2W/2Q$, i.e. unchanged.

(ii) Provided a heat engine is reversible (i.e. no waste of energy) and is working on a Carnot cycle between two given temperatures, it has an efficiency entirely independent of the nature of the 'working substance' and independent of the way in which the external work is performed.

(iii) The efficiency of all reversible engines working on a Carnot cycle between the same two temperatures is the same.

These features caused Kelvin to point out that the ratio Q_1/Q_2 could be used to define a temperature scale independent of the characteristics of any particular thermometric substance. He stipulated that the ratio of any two temperatures is such as to equal the ratio of the heat absorbed to that rejected in a cycle of a hypothetical Carnot engine working between those two temperatures. Thus, in our case,

$$\frac{T_1}{T_2} = \frac{Q_1}{Q_2}, \quad \text{i.e.} \quad \frac{Q_1}{T_1} = \frac{Q_2}{T_2}, \tag{A.2}$$

and therefore

$$\text{Efficiency} \left(= \frac{Q_1 - Q_2}{Q_1} \right) = \frac{T_1 - T_2}{T_1}. \tag{A.3}$$

(The Kelvin 'absolute thermodynamic scale' (or 'work scale') can be shown to be the same as the Ideal Gas Scale but it has wider uses.)

This has an important consequence which leads to the concept known as **entropy**. Consider the two adiabatics in Fig. A.1(b). Q_1 is the quantity of heat required to go from adiabatic 1 to adiabatic 2 along the isothermal at temperature T_1, and Q_2 is the amount required to go from one to the other at temperature T_2. Then in view of Eqn. A.2, we see that whichever isothermal is used to make the change, the value of Q/T is the same. This quantity Q/T that characterises the difference between two adiabatics, is known as the change in entropy. We can write

$$S_1 - S_2 = \frac{Q}{T}, \tag{A.4}$$

where S is the normally-used symbol for entropy.

In the Carnot cycle that we have considered [Fig. A.1(b)], Q_1 is the heat absorbed during B → C. Since Q_2 is the heat *released* during D → A, this should therefore be taken as negative and Eqn. A.2 becomes

$$\frac{Q_1}{T_1} + \frac{Q_2}{T_2} = 0, \qquad (A.5)$$

which means that in a complete reversible cycle, there is no change of entropy.

The same is true for any reversible cycle even if temperature is continually varying. This can be shown by dividing the cycle up into an infinite number of Carnot cycles, leading to the above expression being written as

$$\oint \frac{dQ}{T} = \oint dS = 0. \qquad (A.6)$$

This is regarded as one way of stating the **Second Law of Thermodynamics** since it is based on the assumption of that law.

Note that if an infinitesimal amount of heat is added to a system by a reversible process at a mean temperature T, then the increase in entropy is given by

$$dS = \frac{dQ}{T}. \qquad (A.7)$$

(N.B. dS, unlike dQ, is a perfect differential, being independent, as we have seen, of the path taken.)

$$* \quad * \quad *$$

Entropy is generally regarded to be a difficult concept to grasp and analogies to help have been described. For example, in electrostatics the electrical energy of a charged conductor is proportional to the product of charge and potential:

electrical energy $\propto$ charge $\times$ potential.

Then if we write our definition of entropy as

thermal energy $\propto$ entropy $\times$ temperature,

we see that where temperature corresponds to electric potential, entropy corresponds to electric charge. However, unlike electric charge, entropy is not conserved.

A.2. The Principle of the Increase in Entropy

Unlike the ideal, reversible processes we have considered so far, actual processes are not normally reversible. Every system in the real world, left to itself, changes towards a state of statistical equilibrium. External work is required to drive a system in the opposite direction.

Whilst the **First Law of Thermodynamics** (Joule, 1843) states the conservation of energy, i.e.

$$dQ = dU + dA, \qquad (A.8)$$

where dQ is a small amount of heat given to a system, dU corresponds to the increase in internal energy, and dA is the amount of external work done, the **Second Law** is concerned with the dissipation, or degradation, of energy — it implies the loss of available energy for external use. Clausius pioneered the statement of this in terms of entropy.

Consider a system in which one part at temperature T_1 loses heat to another at a lower temperature T_2 in performing some process in which losses occur. This means that instead of Eqn. A.3, we now have

$$\frac{Q_1 - Q_2}{Q_1} < \frac{T_1 - T_2}{T_1}, \qquad (A.9)$$

i.e. the actual efficiency is less than the ideal. This gives

$$\frac{Q_2}{T_2} - \frac{Q_1}{T_1} > 0. \qquad (A.10)$$

Thus, the entropy of the system has increased, and Clausius regarded it as a general principle. The Second Law of Thermodynamics can then be reformulated to state that the entropy of an isolated system always increases or remains constant.

Of the innumerable thermodynamic relations that follow from the combination of the first and second laws of thermodynamics, the gas laws, and the incorporation of entropy as a thermodynamic function in its own right, there is one we need to note in connection with Planck's work on black-body radiation.

For a system in which no external work is done, Eqn. A.8 becomes

$$dQ = dU,$$

and substituting for dS from Eqn. A.7 we have

$$\frac{dS}{dU} = \frac{1}{T}. \qquad (A.11)$$

A.3. Entropy and Disorder: Planck's Radiation Formula

In the conversion of heat energy into mechanical energy, although the two forms of energy may be equal in amount (**First Law of Thermodynamics**), they are not equivalent in form. Heat energy consists of random motion in a chaotic way, whereas with mechanical work the constituent 'molecules' (or whatever) have to move in an ordered way. The fact that it is easier to produce heat from mechanical work than vice versa means that a system has a tendency to change in the direction of increasing disorder ('probability'): the reverse tendency does not occur unaided. Common experience shows that disordered states are the most probable, and ordered states the least probable.

Boltzmann investigated the **Statistical Mechanics** aspects of thermal equilibrium in great detail. (Maxwell had earlier expressed the view that the second law of thermodynamics is statistical in nature.) Among other things, Boltzmann showed that in a gas in which the molecules do not conform to the **Maxwell–Boltzmann Distribution** (see Appendix B), there is a readjustment towards that distribution. In his 'H-theorem', he gave an expression for the rate at which this approach takes place. It involved a function, H, that decreases as equilibrium is approached. Though seeing this as a link between disorder and entropy, Boltzmann, even in his later major text, *Vorlesungen uber Gastheorie*, did not express it as the well-known

$$S = k \log W, \qquad (A.12)$$

which, however, is named after him. In that text (p133) Boltzmann simply says '...H is proportional to the entropy'.

Planck said that he 'postulated' the above expression in his work on black-body radiation, and we refer to it in more detail in §3.2. As used by Planck, W is the 'probability' in the sense of his calculation of the number of '**complexions**' (see Appendix B, §B.2) of a state whose entropy is S. At the time, k was just a constant but Planck later identified it (in connection with further work on entropy) as the gas constant per mole, but it is nevertheless known as '**Boltzmann's Constant**'.

[If we consider two separate, independent, systems, then their total entropy is the sum of the two entropies. Probability, however, is multiplicative, so the overall probability of the two systems is the product of the separate probabilities. The relationship that meets this requirement is

$$S = k \log W + C, \tag{A.13}$$

where C is a constant. The expression used by Planck left aside the question of an absolute value for entropy, to which he subsequently turned his attention.]

Appendix B

Classical Thermodynamics; Kinetic Theory; Statistical Mechanics; Statistical Thermodynamics

B.1. Introduction

These notes briefly recall the origins and distinguishing features of some related topics that have played a fundamental rôle in the development of quantum mechanics, and are used in numerous places in this book.

Classical thermodynamics is an empirical science based essentially on two laws. The first amounts to a statement of the common experience of the conservation of energy. The second is often expressed as the observation that heat does not flow spontaneously from a cold body to a hotter body. With the concept of entropy (see Appendix A) introduced into thermodynamics by Clausius in 1865, the second law can be reformulated to state that the entropy of an isolated system always increases or remains constant.

Both laws were validated by experimental evidence (as were the laws of classical mechanics which were based on Newton's laws and founded on experience). Mathematical relationships were deduced and the thermodynamic properties of a wide range of physical and chemical systems were dealt with accurately using the thermodynamic variables P, V and T via e.g. **Maxwell's Thermodynamic Relations** (1871).

However, classical thermodynamics is concerned mainly with the state of a system as a whole entity (i.e. bulk properties) and is restricted to systems in equilibrium because it has no means of dealing with the mechanisms of changes or with the structure of matter. For example, it does not provide a satisfactory way of deriving an expression for

239

specific heat, and it cannot account for the unidirectional nature of the second law.

In contrast, the Kinetic Theory of matter was based on the concept of heat as molecular motion. Thus when a gas is confined in a vessel, the collision of the molecules with the container walls is considered. The molecules deliver momentum to the walls, and from Newton's second law it follows that a pressure is exerted on the walls which is the aggregate effect of the impacts. With simplifying assumptions, the pressure can be calculated in terms of the kinetic energy per unit volume.

Kinetic Theory provided a way of deducing the gas laws (Avogadro's Law, Boyle's Law, Dalton's Law) and it equates temperature with the mean kinetic energy of 'molecular' motion. In a gas, the molecules exchange energy by collisions and on the average they will all have the same velocity and kinetic energy. Maxwell in 1860 showed that if in their motions they obey the ordinary laws of mechanics, then the total energy of the system is equally divided on the average among the degrees of freedom. This is Maxwell's **Principle of the Equipartition of Energy.** Of course, the velocities and energies of individual molecules will vary about the mean, and Maxwell derived the distribution of velocities (strictly speeds) about the mean, for any given temperature, for a system of identical but distinguishable 'molecules'. In its simplest form, the **Maxwell Distribution of Velocities** is today written as

$$n(v) = Cv^2 e^{-mv^2/2kT}, \qquad (B.1)$$

where $n(v)$ = number of molecules with velocity v per unit volume, m = molecular mass, C = a constant such that

$$\int_0^\infty n(v)dv$$

equals the total number of molecules per unit volume regardless of speed. The general form is shown in Fig. B.1.

It was the similarity between this distribution of velocities and the black-body radiation spectrum (Fig. 3.2) that was used in the formulation of Wien's Radiation Law (§3.1). In fact the 'Boltzmann

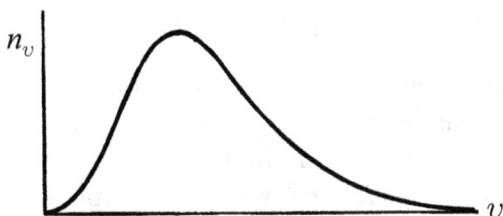

Figure B.1. Maxwell Distribution of Velocities for a given temperature.

constant', k, did not appear as such in Eqn. B.1 until Planck's and Einstein's work on black-body radiation (see also Appendix A, §A.3).

The above distribution of velocities was based on the simplification that the velocity coordinates of the molecules in the gas are independent. Maxwell embarked on a correction for this and it was completed by Boltzmann.

A separate approach to deriving the law was also undertaken at the same time by Boltzmann using the newer methods of Statistical Mechanics emerging from Maxwell's work and developed further by Maxwell, Willard Gibbs, Jeans, and others. It was necessary because Kinetic Theory was seen to be less easily developed to deal with more complicated systems.

Much more widely applicable, **Statistical Mechanics** can be defined as the branch of physics in which statistical methods are applied to the micro-constituents of a system in order to predict its macro-properties. It is as powerful and as widely applicable as thermodynamics but, unlike the latter, it also incorporates the concepts of e.g. molecular interactions. Yet, unlike Kinetic Theory, it does not require detailed knowledge of the forces between molecules, for example. Seemingly, this represents the best of both worlds.

One area of application we are concerned with in this book is the use of statistical mechanics in thermodynamics — **Statistical Thermodynamics**. Here, the thermal parameters of a system are interpreted as an average of the dynamical details of 'molecular' motions, and mathematical deductions can be made concerning e.g. the fluctuations of energy with time, of a molecule about its mean value. '**Fluctuations**' was a topic central to Einstein's studies of radiation (Chap. 4).

B.2.   Maxwell–Boltzmann Distribution

In extending Maxwell's work, Boltzmann considered the question of a polyatomic gas in thermal equilibrium, taking into account possible motions such as rotation and vibrations. The statistical mechanics treatment was very general and widely applicable (to e.g. liquids).

The development by Boltzmann was additionally important because, unlike Maxwell, he started by considering a system of gas molecules having initially an arbitrary distribution of velocities and showed that the distribution always changed towards the one particular steady-state distribution given by Maxwell. Similarly, this occurs for the distribution of energy.

Maxwell–Boltzmann statistics are conveniently stated in terms of the vector quantities momentum, p, and position, q, as 'coordinates' representing the state of a system (see Appendix C concerning the wider use and interpretation of the p, q notation).

For a system of identical, *distinguishable*, 'particles' in thermal equilibrium at temperature T, the statistics tell us that the average number of particles, f, in a particular state (what Planck called a '**complexion**') with energy $E(p, q)$ is given by the 'Maxwell–Boltzmann Distribution Function (or Law)' as

$$f\,[E(p, q)] = Ce^{-E(p,q)/kT}, \tag{B.2}$$

where C is a constant.

The total overall average energy of the system is then given by:

$$\langle E \rangle = \frac{\displaystyle\int E(p, q)e^{-E(p,q)/kT}\,dp\,dq}{\displaystyle\int e^{-E(p,q)/kT}\,dp\,dq}, \tag{B.3}$$

where, in the numerator, we have multiplied each $E(p, q)$ by its probability (cf. Eqn. B.2) and integrated (or summed as appropriate) over all p, q. The denominator is simply for normalisation.

[Maxwell's expression for the distribution of velocities can readily be shown to follow from Eqn. B.3].

Boltzmann's extensive treatment dealt with any number of degrees of freedom and his law proved to hold in terms of 'generalised coordinates' (see Appendix C, §C.3). He had confirmed, refined, and extended Maxwell's work.

Equation (B.2) can be written in the more general form,

$$f(D_i) = Ce^{-D_i/kT}, \tag{B.4}$$

where $f(D_i)$ is the number of particles in a state i with property D_i, as a function of T.

Boltzmann's treatment also gave us the more general **Principle of Equipartition of Energy** with a value of average kinetic energy $KE = \frac{1}{2}kT$ for each degree of freedom. If there is a vibration component, the mean kinetic energy and mean potential energy are shown to be equal, and the total energy per degree of freedom is then kT.

In §4.4, we turn to these matters in connection with Einstein's study of fluctuations which was central to his investigation of thermal radiation.

With the advent of quantum theory, the exactness of these classical statistical mechanics was limited by the **Heisenberg Uncertainty Principle** (§9.2), giving us **Quantum Statistics**.

Appendix C

Phase Space

C.1. Introduction

In classical mechanics, a system in the form of an 'assembly' is described by specifying the coordinates and momenta (used rather than velocities) of all the component 'particles' (atoms, molecules, or other entities). Using rectangular coordinates, this requires three coordinates for each particle, and associated with each coordinate there would be a component of momentum. The name 'phase space', introduced by Willard Gibbs in 1901 in the context of statistical mechanics, was given to the six-dimensional space in which all the spatial coordinates (q_n) and momenta (p_n) for an assembly were plotted as separate variables. (The p, q notation had been introduced by C. G. Jacobi in 1842. See also §7.3.6.)

The concept of a combined position and momentum space enabled statistical mechanics to be developed in a geometrical framework, permitting an easier method of analysis than one wholly abstract in character. To visualise phase space and express it in simple terms, we can just think of it as a two-dimensional plot of p versus q (Fig. C.1). In Appendix B, we saw the Maxwell–Boltzmann distribution function written in this symbolic way Eqn. B.2.

In addition to its rôle in statistical mechanics, phase space has played an important part in various aspects of quantum theory where it was introduced by Planck in his studies (Planck, 1906(a)), particularly on the statistical aspects of entropy, following the outcome of his work on black-body radiation. The following applications of phase space are referred to in various places in this book.

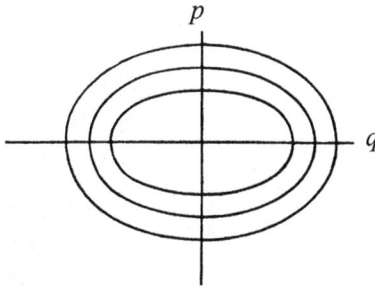

Figure C.1. Phase space.

C.2. Phase Space and Planck's Resonators

An example of Planck's use of phase space was his use of it in considering the simple harmonic oscillator envisaged for his resonators.

Harmonic motion occurs when a body of some kind oscillates, under a restoring force, about an equilibrium position. With simple harmonic motion, a restoring force F acting on a mass m is proportional to its displacement x about the equilibrium, so we have

$$F = -sx, \tag{C.1}$$

where we are here using the familiar 'x', and where s is the 'stiffness constant'. Then by Newton's second law, we have

$$m\ddot{x} = -sx. \tag{C.2}$$

s/m has dimensions of (frequency)^2, so we can write

$$s/m = \omega^2, \quad (\omega = 2\pi\nu), \tag{C.3}$$

and therefore

$$\ddot{x} + \omega^2 x = 0. \tag{C.4}$$

A solution to this is

$$x = a\sin(\omega t + \theta). \tag{C.5}$$

Now momentum is given by

$$p = m\dot{x}$$
$$= m\omega a \cos(\omega t + \theta) \tag{C.6}$$
$$= b\cos(\omega t + \theta),$$

where

$$b = m\omega a.$$

$$\frac{x^2}{a^2} + \frac{p^2}{b^2} = 1. \tag{C.7}$$

Using the p, q notation, we have

$$\frac{q^2}{a^2} + \frac{p^2}{b^2} = 1. \tag{C.8}$$

This relation between the coordinates p and q expresses the state of the oscillator in phase space. The equation describes an ellipse (Fig. C.1), semi-axes a and b, and area given by

$$\pi ab = \pi a^2 m\omega. \tag{C.9}$$

Note that the path round the ellipse is the sequence of points representing the successive states of motion of the oscillator.

In Planck's theory, the interest is in the energy, E, of the resonator. This is given by the sum of the kinetic energy and potential energy at any instant since in the absence of losses this sum is constant throughout an oscillation. We have

$$E = E_{\text{PE}} + E_{\text{KE}}. \tag{C.10}$$

The E_{PE} at displacement x from $x = 0$ is

$$E_{\text{PE}} = \int_0^x sx.dx = \tfrac{1}{2}sx^2. \tag{C.11}$$

The E_{KE} at displacement x from $x = 0$ is

$$E_{\text{KE}} = \tfrac{1}{2}m\dot{x}^2. \tag{C.12}$$

Therefore, the total energy for each oscillation is

$$\begin{aligned} E &= \tfrac{1}{2}sx^2 + \tfrac{1}{2}m\dot{x}^2 \\ &= \tfrac{1}{2}ma^2\omega^2 \\ &= \frac{\pi ma^2\omega}{2\pi/\omega}. \end{aligned} \tag{C.13}$$

$$\therefore E = \text{area of ellipse} \times \nu \tag{C.14}$$

Thus, the energy of a resonator whose state and behaviour is defined by one of these ellipses in phase space is given by the area of that ellipse $\times \nu$.

According to **Planck's quantum condition** (Eqn. 3.37), the energy of a resonator has values restricted to

$$E = nh\nu, \tag{C.15}$$

where $n = 1, 2, 3 \ldots$, so this specifies the possible ellipses (Fig. C.1). The area of the innermost ellipse is h, while that of the second is $2h$, and so on.

When a resonator emits or absorbs energy, its representative p, q point jumps from one ellipse to another, the energy emitted or absorbed being an integral multiple of the energy quantum $h\nu$. In this way, each ellipse corresponds to a stationary state where the resonator can remain without change of energy.

Planck had shown that his quantum condition could therefore be described by the 'phase integral'

$$\oint p.dq = nh, \tag{C.16}$$

where the integration gives the ellipse area, and $n = 1, 2, 3 \ldots$ The above equation, often referred to as **Planck's quantising condition,** is also conventionally written as is also conventionally written as

$$J = nh. \tag{C.17}$$

(In the light of later knowledge, we know that the smallest energy of a resonator is $\frac{1}{2}h\nu$ and not zero as the equation implies.)

With $\oint p \cdot dq$ known as 'action' in Hamilton's mechanics [and originating with de Maupertuis (§7.3.5)] and $h\nu$ known as the 'quantum of energy', Planck named h itself the 'quantum of action'.

As expressed in Eqn. C.16, Planck's quantising condition for his resonators requires that the 'action' over a complete oscillation has a value of nh.

Writing in 1950, W. Wilson regarded Eqn. C.16 as "one of the very greatest of all scientific discoveries of all science ... its discovery places Planck among the half-dozen or so most notable human individuals of all time".

C.3. 'Action' and Electron Orbitals

The quantum condition expressed in Eqn. C.16 is not only identical in principle, as we have just seen, with the quantum conditions used by Planck for the harmonic oscillator, but as we shall now see, it is also consistent with Bohr's orbital theory.

Consider de Broglie's wave model of the electrons in the permitted circular orbitals of Bohr's theory. The orbitals have circumferences equal to a whole number n of de Broglie wavelengths λ (§7.2.2) and therefore the number of de Broglie wavelengths in an element dq of a circumference is dq/λ so that the number in a complete circumference is

$$n = \oint dq/\lambda. \tag{C.18}$$

With de Broglie's equation (Eqn. 7.16),

$$p = h/\lambda, \tag{C.19}$$

this gives

$$\oint p.dq = nh, \tag{C.20}$$

as in Eqn. C.16, but here it can be interpreted as **Bohr's quantising condition** in terms of 'action'.

De Broglie explored these ideas in some detail, demonstrating the analogy between **Maupertuis' Principle** for particles and **Fermat's Principle** for waves, and the relevance to his own work.

W. Wilson, and independently Sommerfeld, generalised the concept in 1915–1916 to be applicable to a great variety of systems of multiple degrees of freedom. They introduced a quantum number for each degree of freedom, and the '**Wilson–Sommerfeld Quantum Conditions**' for orbitals (circular and elliptical) are expressed as

$$\oint p_i.dq_i = n_ih, \tag{C.21}$$

where p_i, q_i are '**generalised coordinates**'.

Sommerfeld applied this to give a complete quantisation of the hydrogen atom (the **Bohr–Sommerfeld theory**). Though extended by

Sommerfeld, Born and Jordan, only problems where variables were separable could, however, be tackled — a limitation overcome by Schrödinger's wave mechanics.

C.4.　Phase Space and Statistics

To illustrate the use of phase space in statistics, we can first look at its use in Boltzmann's classical statistics of a perfect (ideal) gas. The gas molecules are assumed to be monatomic, very small, identical, perfectly elastic spheres, with no mutual interactions other than collisions. Each has its p, q represented by a phase point in a six-dimensional phase space. If there are N molecules in the system, then for a given 'microstate', the N phase points of the molecules are distributed among various '**phase cells**', which we will assume, for the moment, to be infinitesimally small. For a given microstate, let there be phase points in cell.

a_1	phase points in cell	1
a_2	"	2
$\vdots$		$\vdots$
a_n	"	n

In a gas, there are exchanges of position and momentum between molecules for which the numerical occupation of cells (and the total energy) is unchanged. From a macroscopic point of view, the situation does not change. The number of these different microstates in a single macroscopic state is given by the number of different ways in which the points may be interchanged among the cells whilst the number of points in each cell remains unchanged. This number is

$$W = \frac{N!}{a_1!a_2!a_3!\ldots a_N!}. \tag{C.22}$$

The 'probability' of a given macrostate is naturally associated with the value of W. The macrostate with the largest W is the one most likely to endure and it comprises the state of '**statistical equilibrium**'.

As this is the state to which the gas tends to pass, it is assumed to be the state of the highest entropy.

According to the usual definition of '**probability**', its value in the above example would be W/D where D is the total number of different distributions comprising all possible macrostates. Also, since entropies are additive and probabilities are multiplicative, the relationship between entropy and probability can be expressed as a proportionality,

$$S = A \log(W/D), \qquad (C.23)$$

where A is a constant to be determined. (In Planck's black-body radiation studies, he had simply postulated (§3.2.3) $S = k \log W$ and as we know, later identified k with the gas constant.)

When Planck had found it necessary in 1900 to quantise the possible energies of his resonators, this could now be interpreted in this later work of 1912 as putting a finite size to the cells of phase space: there were restrictions to p, q . He assigned a value to the size of the phase cells in the following way.

Let the cell dimensions (bearing in mind that this is a six-dimensional space) be

$$\Delta q_x, \Delta q_y, \Delta q_z, \Delta p_x, \Delta p_y, \Delta p_z.$$

Now Planck's work on blackbody radiation had established the existence of the '**quantum of action**', h (§3.2.3), and since $\Delta p \Delta q$ has the dimensions of '**action**', which can be expressed as momentum × distance (§7.3.5), he considered it plausible to set

$$\Delta q_x \Delta p_x = \Delta q_y \Delta p_y = \Delta q_z \Delta p_z = h. \qquad (C.24)$$

The volume of each cell in the phase space for black-body radiation was then h^3.

Planck used this in his continuing interest in the statistical aspects of entropy, and his method of calculating the entropy of a gas drew attention in 1921 to the close relationship between thermodynamics (the Third Law) and quantum theory.

This use of phase space was also adopted in 1924 by Bose and Einstein who made major advances in quantum statistics, starting with Bose's way of establishing Planck's radiation formula without recourse to the use of electromagnetic theory [as in the derivation of Planck's key starting point, viz. Eqn. 3.9]. Extended to a 'gas theory', their work also provided an important inspirational background to Schrödinger's way of formulating his wave mechanics (§7.3.2).

Another point of relevance and interest for us is that the above choice of cell size, involving h, was later further vindicated by being seen as a consequence of Heisenberg's Uncertainty Principle (§9.2).

Appendix D

A Note on Rayleigh's Radiation Formula

In 1900, Lord Rayleigh published an approach to black-body radiation (with developments by Sir James Jeans and others in the following few years) using the view of thermal radiation as consisting of vibrations governed by the equations of Maxwell's Electromagnetic Theory and Newtonian dynamical principles, in an all-pervading aether. On that basis, a black-body chamber would be filled with such vibrations and it was shown that the number of independent modes of vibration, per unit volume, of frequency between v and $v + dv$ is

$$\frac{8\pi v^2 dv}{c^3} = \frac{8\pi}{\lambda^4} d\lambda.$$

It should be noted that Rayleigh obtained the constant term in the above by a method different from that used by Planck (Eqn. 3.9). He dealt directly with the standing waves he visualised to exist in the radiation field and did not involve the idea of equilibrium with the material comprising the cavity walls — which was Planck's approach.

Assuming that the **Principle of the Equipartition of Energy** applies to vibrations in the aether in equilibrium at absolute temperature T, the energy of the radiation between the above limits is therefore

$$\rho(v, T) = \frac{8\pi v^2}{c^3} kTdv = \frac{8\pi}{\lambda^4} kTd\lambda, \tag{D.1}$$

and is known as the **Rayleigh (or Rayleigh–Jeans) Radiation Formula (or Law).**[h]

---

[h] Rayleigh's range of scientific activities was prodigous. His Nobel Prize for physics in 1904 was awarded for work on gases — including the discovery of argon.

The linear relationship with T proved to be in good agreement with experiment at long wavelengths and it had the advantage of not involving Planck's concept of resonators. However, it failed to show the well-established maxima (Fig. 3.2) and was in serious conflict at short wavelengths because according to the formula, the energy of the radiation would tend to infinity — what became referred to as the 'UV catastrophe'. To suppress the latter, an ad hoc exponential term was added. Jeans believed that the failure of their approach had been due to the absence of true thermal equilibrium in the experimental radiation, not to lack of soundness of the equipartition theorem (Klein, 1966).

In 1910, Debye used this approach but obtained the correct (Planck) formula by introducing quantisation to replace the idea of Equipartition of Energy (see Appendix E, §E.2).

Appendix E

Debye: Specific Heat Theory of Solids and Derivation of Planck's Radiation Formula

During 1910–1912, Debye made two important contributions to the developing quantum story. They were both based on applying quantisation to vibrational spectra. For that reason, we bring them together here, and in an appendix because they are referred to in several chapters. One was a development of Einstein's specific heat theory and the other was an independent derivation of Planck's radiation formula.

E.1. Specific Heat of Solids

Dulong and Petit's law of 1819 had stated that the specific heat per mole for all solids is approximately 6 cals/degree ($25 \, \mathrm{J \, mol^{-1} \, K^{-1}}$). We saw in §3.1 that it was based on atoms being assigned a mean energy of $3kT$ in accordance with the **Principle of the Equipartition of Energy**. Whilst in good agreement with experiment at room temperatures, it failed at low temperatures where specific heats diminish (Fig. 3.1).

Then Einstein, in 1907, showed that quantising the energy in accordance with Planck's theory gave better agreement with experiment (§4.5). However, Einstein used the simple model of a solid in which all atoms oscillate isotropically, independently, and with the same frequency. Further refinement was needed.

Debye took into account the spectrum of atomic oscillation frequencies that exist in a solid.

In the normal theory of elasticity, the vibrations of the atoms in, say, a crystal are treated as vibrational modes of the whole crystal. As a simplification, Debye took the example of a monatomic crystal and 'replaced' the vibrational energy spectrum of the atoms by the spectrum of elastic vibrations in the crystal. Determining the spectrum of vibrations in a body regarded in elasticity theory as continuous, is analogous to the familiar one-dimensional case of a vibrating string. In the simplest case, it gives the number of vibrational modes, per unit volume, in the frequency range $v \to v + dv$, as

$$\frac{4\pi v^2}{c^3} dv,$$

where c is the velocity of elastic sound waves in the solid.

Returning to this as representing the atomic vibrational spectrum, and using Planck's expression (Eqn. 3.33) for the mean equilibrium energy for a given frequency v, we obtain the total mean vibrational energy of the atoms in unit volume of the solid as

$$\int_0^{v_{max}} \frac{hv}{e^{hv/kT} - 1} \cdot \frac{4\pi v^2}{c^3} dv,$$

where v_{max} is determined by the dimensions of the solid.

From this, Debye calculated the specific heat and it gave a better match with experimental data than had Einstein's model based on independent oscillators all having the same frequency (Fig. 3.1).

Debye's treatment was further improved by Born and Kármán, and others, in the light of the developing knowledge of the structure of solids following the Laue experiment in 1912.

E.2. Debye's Derivation of Planck's Radiation Formula

Debye adopted a similar approach to the above to derive Planck's radiation formula. His successful method was an important background influence when Schrödinger was conceiving his wave mechanics (Chap. 7).

Unlike Planck, and like Rayleigh (see Appendix D), he considered the actual radiation in a black-body cavity. The number of independent modes for unit volume, of frequency between v and $v + dv$ was then

$$2 \times \frac{4\pi v^2}{c^3} dv,$$

with the factor 2 because, with radiation, there are two independent vibrations associated with polarisation.

Unlike Rayleigh, however, who had invoked the Principle of Equipartition of Energy, Debye introduced quantisation by limiting the admissible energies to $nh\nu$, $n = 0, 1, 2, \ldots$

Each energy was weighted according to the Boltzmann factor (see Appendix B, Eqn. B.2) whence the average energy for frequency ν is

$$E(\nu, T) = \frac{\sum\limits_{n} nh\nu e^{-nh\nu/kT}}{\sum\limits_{n} e^{-nh\nu/kT}}.$$

The numerator is

$$h\nu e^{-h\nu/kT} + 2h\nu e^{-2h\nu/kT} + 3h\nu e^{-3h\nu/kT} + \ldots,$$

and the denominator (for normalisation) is

$$1 + e^{-h\nu/kT} + e^{-2h\nu/kT} + e^{-3h\nu/kT} + \ldots.$$

Using the binomial theorem for both gives

$$E(\nu, T) = \frac{h\nu}{e^{h\nu} - 1}.$$

Multiplying this by the above number of modes for frequency ν, we obtain the radiation spectral distribution

$$\rho(\nu, T) = \frac{8\pi\nu^2}{c^3} \frac{h\nu}{e^{h\nu/kT} - 1},$$

exactly as Planck had obtained (Eqn. 3.34) from totally different premises.

Appendix F

The Photoelectric Effect

The discovery of the photoelectric effect is attributed to Hertz in 1887 since it emerged from studies inspired by his observations in the course of experimental work on electric discharges across spark gaps. Hallwachs in the following year, Stoletov in 1890, J. J. Thomson in 1899, and Lenard in 1902, progressively established that the effect is due to the emission of electrons from a metal surface when light of sufficiently high energy (e.g. ultraviolet) falls upon it.

Briefly, in studies of the effect, emitted 'photoelectrons' were attracted to an electrode by an applied voltage, and the photoelectric current measured. If the voltage was increased sufficiently, the current levelled off at a value at which all the emitted electrons were collected [Fig. F.1(a)]. The voltage was then reduced until the current dropped to zero. The voltage at which that occured gave the maximum energy of the photoelectrons, their '**stopping power**', for a given frequency of illumination.

Two unaccountable features were observed: (i) the energy of the photoelectrons was independent of the intensity of the incident light, and (ii) whilst the photoelectron energy depended on the frequency of the illumination, as one would expect, it was found that at frequencies below a certain threshold frequency, ν_0, characteristic of the metal, no electrons were emitted [Fig. F.1(b)].

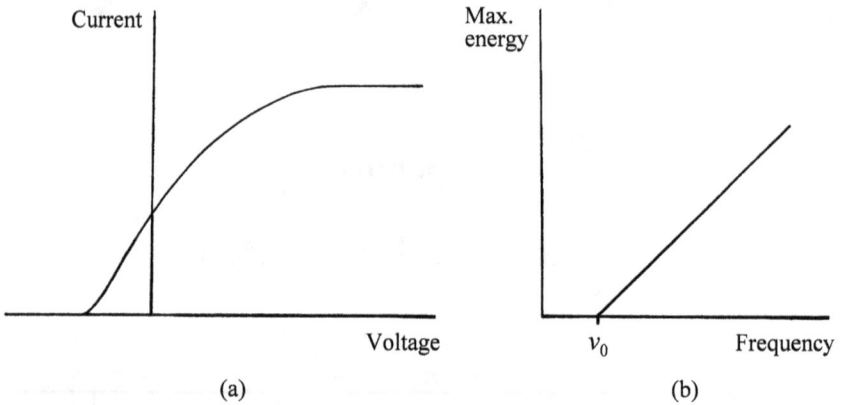

Figure F.1. Photoelectric effect.

Einstein gave an explanation of these effects in the second part of the 1905 paper in which he introduced the concept of light quanta (§4.3) (Einstein, 1905(a)). His work on the photoelectric effect was confirmed experimentally by Millikan in 1916.

Appendix G

The General Wave Equation; Wave Groups; Dispersion

G.1. The General Wave Equation

The 'general wave equation', derived in elementary texts for a simple harmonic transverse plane wave travelling along a stretched string in the x direction, is commonly expressed as

$$\frac{\partial^2 y}{\partial x^2} = \frac{1}{u^2}\frac{\partial^2 y}{\partial t^2},\tag{G.1}$$

where $y =$ displacement in the plane of the wave,

and $u =$ **wave (phase) velocity.**

One solution often given is

$$y = A\cos(\omega t - kx),\tag{G.2}$$

using $\omega = 2\pi v$ (angular frequency),

and $k = 2\pi/\lambda$ (angular wave number).

$$\therefore u = \omega/k$$

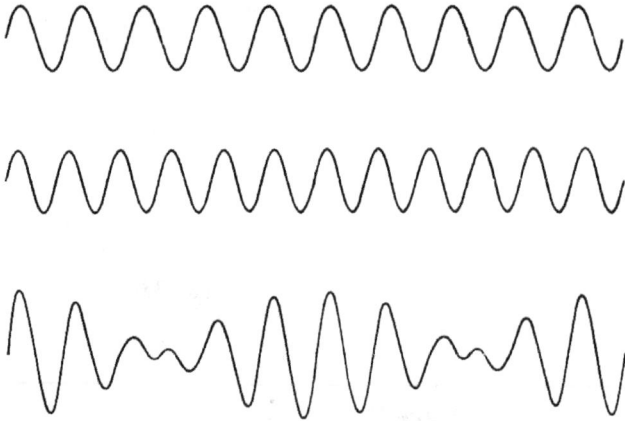

Figure G.1. Lower wave is the summation of the two above.

The general solution is then

$$y = Ae^{-\omega(t-x/u)}. \tag{G.3}$$

G.2. Wave Groups

As a simple illustration of wave groups, we can consider the wave groups formed by the addition of two waves of the type specified in Eqn. G.2, with the same amplitudes but slightly different frequencies:

$$
\begin{aligned}
y_1 + y_2 &= A \cos(\omega_1 t - k_1 x) + A \cos(\omega_2 t - k_2 x) \\
&= 2A \cos\left[\left(\frac{\omega_1 - \omega_2}{2}\right) t - \left(\frac{k_1 - k_2}{2}\right) x\right] \\
&\quad \times \cos\left[\left(\frac{\omega_1 + \omega_2}{2}\right) t - \left(\frac{k_1 + k_2}{2}\right) x\right].
\end{aligned} \tag{G.4}
$$

This is a wave system (Fig. G.1) with a frequency $(\omega_1 + \omega_2)/2$ close to those of the components, but with an amplitude $2A$ modulated in space and time by a slowly varying envelope of angular frequency $(\omega_1 - \omega_2)/2$ and angular wave number $(k_1 - k_2)/2$.

The velocity of this envelope is the '**group velocity**', u_g, and is therefore given by

$$u_g = \frac{\omega_1 - \omega_2}{k_1 - k_2}$$

$$= \frac{d\omega}{dk} \quad \text{in general}$$

$$= \frac{dv}{d(1/\lambda)}$$

or

$$= \frac{dv}{dv'}, \tag{G.5}$$

where v' is used to denote **wave number**.

G.3. Dispersion

When phase velocity, u, is wavelength dependent, i.e. ω/k is not constant, there is said to be **dispersion**. Referring back to Eqn. G.5 we have

$$u_g = \frac{d\omega}{dk}$$

$$= \frac{d(ku)}{dk}$$

$$= u + k\frac{du}{dk},$$

or, as often expressed,

$$u_g = u - \lambda\frac{du}{d\lambda}. \tag{G.6}$$

Thus, if there is no dispersion, the group velocity is the same as the velocity of the constituent wave motion. However, in general the two are not the same.

When $du/d\lambda$ is positive, there is said to be **normal dispersion**, and the group travels slower than the constituent wave motions. When $du/d\lambda$ is negative, there is said to be **anomalous dispersion**.

An important feature of group velocity is that it is the velocity with which energy is propagated.

Appendix H

The Harmonic and Anharmonic Oscillator

H.1. The Simple (Linear) Harmonic Oscillator

In Appendix C, we noted some details of the simple harmonic oscillator used by Planck as a model for the 'resonators' in his black-body radiation studies (Chap. 3). Here, we reiterate that in order to extend it to deal with the anharmonic oscillator in the way needed in Chap. 8 in connection with matrix mechanics.

Harmonic motion occurs when a system vibrates about an equilibrium configuration. It requires the presence of a restoring force, and we can first consider the case when the restoring force F is proportional to displacement x (using one dimension for simplicity) about the equilibrium position i.e.,

$$F = -sx, \tag{H.1}$$

where s is a constant (the 'stiffness constant'). This is **Hooke's law**, and with Newton's second law of motion, we have the equation of motion,

$$m\ddot{x} = -sx,$$

or

$$\ddot{x} + \frac{s}{m}x = 0, \tag{H.2}$$

where m is the mass of the entity subjected to the force. s/m has the dimensions of (frequency)^2 ($MLT^{-2}/ML = T^{-2}$), so the above equation can be written [as in Eqn. C.4],

$$\ddot{x} + \omega^2 x = 0, \qquad (H.3)$$

where ω = angular frequency (radians/s).

A system obeying this is known as a simple (or linear) harmonic oscillator. A simple solution is, for example,

$$x = a \cos \omega t$$

or

$$= a \cos 2\pi v t, \qquad (H.4)$$

where a = amplitude, v = frequency (cycles/s).

However, this restricts the oscillation to be at its full amplitude at $t = 0$. More generally, we can write

$$x = a \cos 2\pi v t + b \sin 2\pi v t, \qquad (H.5)$$

where a suitable choice of a and b determines the value of x at $t = 0$. The amplitude at any time is given by $\sqrt{a^2 + b^2}$.

The importance of the harmonic oscillator in quantum physics lies not in an assumption of the general validity of **Hooke's law** but in the fact that restoring forces do conform to it for small displacements. [This is easily shown by expressing any force, $F(x)$, as a **Maclaurin's series** about the equilibrium position $x = 0$. Thus,

$$F(x) = F_{x=0} + \left(\frac{dF}{dx}\right)_{x=0} x + \frac{1}{2}\left(\frac{d^2F}{dx^2}\right)_{x=0} x^2 + \dots, \qquad (H.6)$$

where $F_{x=0} = 0$ since $x = 0$ is the equilibrium position. For small x, terms in x^2 and higher are small compared with x, leaving

$$F(x) = \left(\frac{dF}{dx}\right)_{x=0} x \qquad (H.7)$$

which is Hooke's law, where $\left(\frac{dF}{dx}\right)_{x=0}$ is negative for a **restoring force**. Thus, we have the conclusion that all oscillations are simple harmonic when their amplitudes are sufficiently small.]

We can note that the potential energy of a mass m when it is displaced from its equilibrium position to a distance x is given by the work done to get there, i.e.,

$$PE = -\int_0^x F dx = s \int_0^x x dx$$

$$= \frac{1}{2} s x^2.$$

An alternative, equivalent, solution to Eqn. H.3 can more usefully be expressed for our purposes using complex exponential notation,

$$x = q_1 e^{2\pi i v t} + q_2 e^{-2\pi i v t} \tag{H.8}$$

where, for x real, the complex constants q_1 and q_2 are given by

$$q_1 = \frac{a - ib}{2}, \quad q_2 = \frac{a + ib}{2}. \tag{H.9}$$

q_1, q_{-1} are complex conjugates and we denote this in the conventional way as follows:

$$q_1 = q_{-1}^*, \quad q_1^* = q_{-1} \tag{H.10}$$

and intensity (amplitude squared) is proportional to $q_1 q^*$.
[Equation H.8 becomes

$$x = \left(\frac{a - ib}{2}\right)(\cos 2\pi v t + i \sin 2\pi v t)$$

$$+ \left(\frac{a + ib}{2}\right)(\cos 2\pi v t - i \sin 2\pi v t)$$

$$= a \cos 2\pi v t + b \sin 2\pi v t$$

as in Eqn. H.5]
To allow for any point on the line of oscillation to be taken as the origin, not necessarily $x = 0$ as in Eqn. H.8, we rewrite the latter as

$$q = q_0 + q_1 e^{2\pi i v t} + q_{-1} e^{-2\pi i v t}, \tag{H.11}$$

where q_0 is real and is the distance of the chosen centre of oscillation from the origin $x = 0$.

As the amplitude of oscillation of such a linear harmonic oscillator increases, the energy of the motion increases accordingly, but the frequency of oscillation is a characteristic of the system and remains the same. In the case of a *freely* oscillating electron, the laws of classical electrodynamics require that the radiation emitted is monochromatic, with frequency exactly that of the oscillating electron.

H.2. The Anharmonic (Non-Linear) Oscillator

Unlike the example above, of a *freely* oscillating electron, a change in the energy of an *orbital* electron occurs when there is a change of orbital and this is associated with a change in frequency. The model for dealing with this is the **anharmonic oscillator**. Equation H.3 then has an added term (or terms) such as

$$\ddot{x} + \omega^2 x + \mu x^2 = 0, \tag{H.12}$$

where μ is a constant.

Fourier's theorem shows that for a frequency of oscillation v, the distance q of the oscillating mass from the chosen origin in the line of oscillation is given by the sum of a converging infinite series of harmonics of v. Instead of Eqn. H.11, we have

$$q = q_0 + q_1 e^{2\pi i v t} + q_2 e^{2\pi i (2v)t} + q_3 e^{2\pi i (3v)t} + \cdots$$
$$+ q_{-1} e^{-2\pi i v t} + q_{-2} e^{-2\pi i (2v)t} + q_{-3} e^{2\pi i (3v)t} + \cdots \tag{H.13}$$

i.e.

$$q = \sum_{-\infty}^{+\infty} q_s e^{2\pi i s v t}, \tag{H.14}$$

where $s = 0, \pm 1, \pm 2, \dots$.

q_0 (i.e. $q_{s=0}$) is a constant equal to the average distance of the oscillating mass from the origin ($x = 0$). Otherwise, for q real, we have the conjugates

$$q_s = q^*_{-s},$$

and

$$q^*_s = q_{-s},$$

(H.15)

with energy proportional to $\sqrt{q_s q^*_s}$.

For any particular energy, say $E^{(1)}$, we can denote the fundamental frequency of oscillation as $v^{(1)}$, with harmonics $2v^{(1)}, 3v^{(1)}, \ldots$. Similarly, those applies for other energies $E^{(2)}, E^{(3)}, \ldots$, etc.

There is a special way of tabulating the Fourier terms for various energies of oscillation. This is shown below:

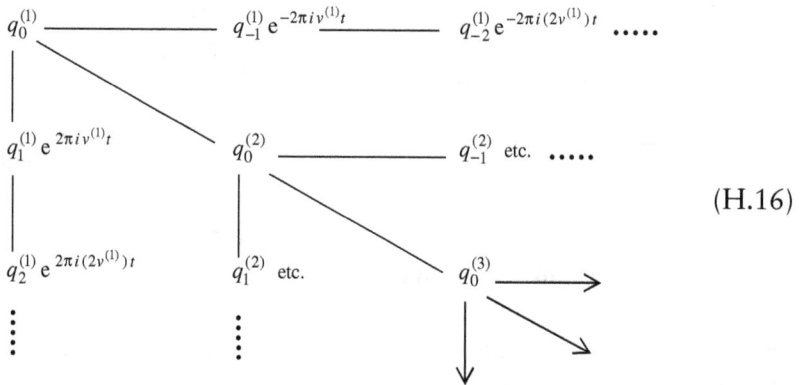

(H.16)

Here, the constant terms $q_0^{(n)}$ of the Fourier series for oscillation energy $E^{(n)}$ are placed along a diagonal, and the pairs of associated Fourier terms for the harmonics (cf. Eqn. H.13) are listed down from the $q_0^{(n)}$ and to the right of it.

This is the situation for a **classical anharmonic oscillator**. However, these frequencies and amplitudes would not be correct for quantum transition energies and Heisenberg's aim was to retain the general form of this array but to replace the classical magnitudes by the corrected quantum magnitudes (Chap. 8).

Appendix I

Chronology of Main Developments Leading to the Copenhagen Interpretation

1900

Planck postulated discrete 'energy elements' (later 'quanta') to explain the production of black-body radiation, but he firmly believed that the radiation itself is transmitted as wavemotion in accord with Maxwell's classical **Electromagnetic Theory** (Chap. 3).

1905

Of Einstein's major contributions in this year one was his **Special Theory of Relativity** (Chap. 2). In another he developed and justified the concept of light being transmitted as discrete 'energy quanta' (Chap. 4). He showed that this successfully explained the **photoelectric effect**. Phenomena such as diffraction were to be regarded as time averages rather than instantaneous values.

1909

Einstein's work now suggested a need for a 'fusion' of wave-quantum interpretations of light (§4.2).

1913

Bohr showed that quantisation explained atomic orbital structure and the details of simple spectra. He showed there is a region of scale where the new quantum laws merge into classical laws — his **Correspondence Principle**. He was to remain forever a staunch believer in quantum jumps, such as those occuring in atomic orbital transitions, but like Planck he was reluctant to accept the concept of quanta in the actual transmission of the radiation resulting from those transitions (Chap. 5).

1916–1917

Einstein established more explicitly the association of momentum with light quanta (§4.2).

1923

The **Compton Effect** confirmed the reality of 'light quanta' and their having momentum (§4.2). De Broglie, noting Einstein's duality concept in connection with radiation, proposed that such duality might also apply to matter particles. This was confirmed experimentally, but the meaning of his '**pilot waves**' accompanying particles was not satisfactory. A more satisfactory '**wave mechanics**' was developed by Schrödinger in 1926 (Chap. 7).

1924

The first thoughts were recorded about the problem of the wave-particle duality of radiation. Published by Bohr, Kramers and J. Slater, it is known as the '**BKS Suggestion**' (mentioned briefly in §5.3). It stemmed largely from Bohr's reluctance to accept the concept of the photon in the actual transmission of light and attempted to account for particle behaviour (as needed to explain e.g. the photoelectric effect) without recourse to the concept of light quanta. The reasoning was laboured, involving Slater's idea of (i) a '**virtual radiation field**', (ii) the notion of conservation of energy and momentum as only statistical and not applicable at the elementary level, and (iii) loss of causality. Although the Compton Effect results demolished the BKS suggestion, it was, as Heisenberg reflected in 1955, an important first step after 20 years, to resolve the paradoxes left by Planck's theory and Einstein's work on light quanta which had introduced probability as a 'new kind of objective physical reality'.

At this time, Bohr was developing his **principle of complementarity** as a way of dealing with **wave–particle duality** (Chap. 9).

1925

Einstein's theoretical work gave further, independent, support to de Broglie's proposal for a wave–particle duality of matter (Chap. 4).

Also in this year, Heisenberg published a 'new mechanics' to be interpreted almost immediately by Born and Dirac as based on matrix algebra, hence 'matrix mechanics' (Chap. 8). Like Bohr, he was committed to the particle aspect of quantum theory, with its discontinuities and quantum jumps. He argued that wave mechanics could not explain quantum phenomena such as Planck's radiation formula, the Compton Effect, or Einstein's work on spectroscopic stationary states: these all required the mechanism of discontinuity and quantum jumps. Unlike Bohr, however, Heisenberg believed in dealing only with observables, and as we see in Chap. 8, this was the basis on which he developed his 'new mechanics'. After the initial identification of it as a matrix mechanics, it was put on a firm basis at the end of 1925 by the famous '**three-man paper**' (*dreimannerarbeit*), viz. Born, Heisenberg, and Jordan (1926).

Dirac was shown proof-sheets of Heisenberg's paper, and Dirac's biographer, Helge Kragh, recounts how Dirac found that the treatment was not only complicated and unclear but also did not take relativity into account. He tried several abortive approaches but made important progress when he saw a resemblance between Heisenberg's non-commuting variables and non-commuting Poisson bracket algebra which can be used to formulate **Hamiltonian dynamics**. After learning more about that algebra, he 'could now proceed to formulate the fundamental laws of quantum dynamics from classical mechanics in its Poisson bracket formulation'. His famous paper, written in late 1925, 'The fundamental equations of quantum mechanics', was published in 1926.

1926

Almost simultaneously with Heisenberg's work on a new mechanics, Schrödinger developed de Broglie's idea into a more rigorous and usable 'wave mechanics' in which waves were the fundamental constituents of matter. For example, an electron was to be regarded as a wavegroup, consisting of a number of amplitude functions, each a solution of his wave equation. Quantum jumps were then visualised as changes in harmonics. He was the first to establish the relationship between his wave mechanics and matrix mechanics. He did this by showing that Heisenberg's matrices could be constructed from the solutions of his differential equations, and vice versa. Schrödinger thought he had eliminated Bohr's quantum jumps, but his model failed to explain experimental observations such as the tracks seen with the Wilson cloud chamber.

Much to Schrödinger's dismay, Born and his assistant Jordan immediately showed that Schrödinger's waves could better be regarded as waves of probability associated with, but not replacing, particles of matter. In this scenario, the determinism of classical physics was at an end. Schrödinger, however, continued to believe firmly in the continuity of matter. Einstein was enthusiastic about Schrödinger's wave mechanics (with which he had a link via Schrödinger's use of his $E = h\nu$) but was always sceptical about Born's statistical interpretation. He could not accept the implications of a physical world that is fundamentally indeterminate (cf. his remark often quoted as '**God does not play dice**'). Also he had a long-running 'conflict' (his word) with Bohr over the latter's reluctance to accept the photon concept of light transmission, despite the latter's introduction of the quantum in explaining the generation of optical linespectra. Einstein's belief that underlying quantum theory there must be a deeper, deterministic theory of nature, lost favour in the face of the huge success of quantum theory — causing most physicists to now believe that the ultimate physical theory is quantum mechanical and that nature is fundamentally random.

*   *   *

Such was the general outline of the state of affairs at the time of the historic meeting that took place in Copenhagen in the autumn of 1926 (Chap. 9).

Appendix J

Planck Units

In a lecture entitled 'On the physical units of Nature' given to the British Association in 1874, G. Johnstone Stoney (who coined the names 'electron' (see p. 48) and 'ultraviolet') was the first to conceive the idea of natural units, or absolute units, i.e. units based on quantities that are regarded as natural, universal constants, as opposed to being human-based definitions. In the lecture, published seven years later, he introduced such units in terms of c, G, and the charge, e, of the electron (Stoney, 1881; Ray, 1981).

Then, in 1899, when Planck was discovering his famous constant, h, perhaps unaware of Stoney's suggestion, he became interested in the prospect of '**natural units**' (Planck, 1899). As Abraham Pais colourfully put it, Planck "concluded the nineteenth century part of his career with an invocation of the absolute" (Pais, 1991).

By dimensional analysis, Planck found that with different combinations of a constant in Wien's law (Eqn. 3.8), the velocity of light (c) *in vacuo*, and Newton's gravitation constant (G), he could derive units of length, mass, time, etc., that were thereby 'universal' (Planck, 1900). Using experimentally obtained values for those constants, he originally obtained the following values:

Length $\qquad\qquad \sqrt{\dfrac{\beta G}{c^3}} = 4.13 \times 10^{-33} \text{ cm}$

Mass $\qquad\qquad \sqrt{\dfrac{\beta c}{G}} = 5.56 \times 10^{-5} \text{ g}$

Time $$\sqrt{\frac{\beta G}{c^5}} = 1.38 \times 10^{-43}\,\text{s}$$

With Planck's constant (h) established, Planck was then able to express them without recourse to Wien's law (Planck, 1906(b)), and with the later-adopted 'reduced' Planck constant, $\hbar = h/2\pi$, the values employed today are expressed as:

Length $$L_p = \sqrt{\frac{\hbar G}{c^3}} \approx 1.6 \times 10^{-33}\,\text{cm}$$

Mass $$M_p = \sqrt{\frac{\hbar c}{G}} \approx 2.2 \times 10^{-5}\,\text{g}$$

Time $$T_p = \sqrt{\frac{\hbar G}{c^5}} \approx 5.4 \times 10^{-44}\,\text{s}$$

Similar expressions for energy, temperature, etc. are in use.

In general, there is no reason why dimensional analysis results are necessarily meaningful, and they need to be interpreted carefully. For example, stating a length as so many L_p units may be universally convenient but it may have no value or meaning beyond that.

However, since c, G, and h, each play fundamental rôles it was realised that a 'Planck Scale' based solely on them may be meaningful. Thus, c links space with time in Special Relativity, G links space-time curvature and energy-momentum density in General Relativity, and h is the very foundation of Quantum Theory. It is interesting to note that using various fundamental relations of this sort (mostly not known when Planck was using dimensional analysis) Wayne Saslow derived, heuristically, the same expression as Planck for L_p (Saslow, 1998).

Immediate reactions are, of course, to the magnitudes of the Planck Units. The Planck unit of mass is ca. 10^{19} times the proton mass and this draws attention to the lack of knowledge about the origin of mass. However, most units are unimaginably small. For example, the Planck unit of length is ca. 10^{-20} of the size of the proton. This is clearly well into the realm of the Uncertainty Principle with its inherent quantum fluctuations to which the smoothness of G in the expressions for Planck units must surely then also be subject. Hence the search for a quantum theory of gravity that unites Quantum Mechanics with General Relativity.

Not surprisingly, for reasons such as the above example, the Planck scale is now brought into a very wide range of topics in cosmology, quantum field theory and quantum gravity, the origin of mass, particle physics, string theory, M-theory, etc. Of course, there are potentially misleading dangers in this: with 'quantum gravity', for example, it may be that its basis is not connected with all of the physical constants that combine to specify Planck units, but with physical mechanisms at present unknown.

Further Reading

Gearhart C. A. 2002 — see Bibliography.
Greene B. 2004.
   The Fabric of the Cosmos: Space, Time, and the Texture of Reality.
   London: Penguin Press Science.
Majid S. (ed.) 2008.
   On Space and Time. Cambridge: Cambridge University Press.
Meschini D. 2007.
   Planck-scale Physics: Facts and Beliefs. *Foundations of Science* **12**:
   277–294.
Saslow W. M. 1998.
   A Physical Interpretation of the Planck Length. *Eur. J. Phys.* **19**: 313.
Silvaram C. 1986.
   The Planck Length as a Cosmological Constraint. *Astrophysics and
   Space Science.* **127**: 133–137.
Wilczek F. 2001–2002.
   Scaling Mount Planck. Part 1: June 2001, Part 2: November 2001, Part 3:
   August 2002. *Physics Today.* New York: American Institute of Physics.
Wilczek F. 2008.
   The lightness of being: Mass, Ether, and the Unification of Forces.
   New York: Basic Books.

Appendix K

Biographical Notes: The Central Characters

(Based largely on R. L. Weber *Pioneers of Science*)

Planck, Max Karl Ernst Ludwig (1858–1947)

Nobel Prize 1918

Max Planck was born in Kiel, Prussia, the son of a professor of law. The family moved to Munich in 1867, and Planck studied at the university there from 1874 to 1877. He received his PhD at the University of Berlin in 1879 under Helmholtz and Kirchhoff, and in 1889 became Kirchhoff's successor, remaining there until he retired in 1926.

Planck's early research was in thermodynamics, followed by work in mechanics, optical and electrical problems, and the radiation of heat, the latter leading to the quantum theory.

AIP Emilio Segre Visual Archives.

Planck had a son and two daughters by his first marriage in 1887, and one son by a second marriage in 1911. He suffered many personal misfortunes, and tragedies including the loss of his children — a son in the First World War, his twin daughters both died in childbirth, and his eldest son was executed in 1945 for suspected involvement in a plan to overthrow the Nazi regime. His home and library were destroyed in an air-raid on Berlin, and he spent his last days in the home of his grand-niece in Göttingen where he died in 1947 and was buried in Göttingen's Stadtfriedhof.

Einstein, Albert (1879–1955)

Nobel Prize 1921

Hebrew University of Jerusalem Albert Einstein Archives, courtesy of AIP Emilio Segre Visual Archives.

Albert Einstein was born in Ulm, Germany, in 1879. In 1885 the family moved to Munich where Einstein's father had a small, financially unsuccessful, electrical and engineering firm. After attending a school in Aarau, Switzerland, for a year, Einstein was admitted to the Swiss Federal Polytechnic School in Zurich which he attended from 1895 to 1900. He married Mileva Maric, a mathematician, in 1901, and they had two sons, Albert and Eduard.

In 1902, Einstein became a technical assistant at the Patent Office in Bern. He received his PhD from the University of Zurich in 1905 — the year that saw the publication of the history-making three papers.

In 1909, Einstein left the Patent Office in Bern to become 'professor extraordinary' at the University of Zurich, then moving in 1913 to become the first director of the Kaiser Wilhelm Physical Institute in Berlin. At this time he separated from his wife, and later married his cousin Elsa.

From 1914 to 1933, Einstein was professor of physics at the University of Berlin, during which time he travelled widely abroad. In 1933, he was appointed life member of the Institute for Advanced Study in Princeton, becoming a US citizen in 1940.

Having made many major contributions to physics, referred to in numerous places in this book, by 1953, Einstein had turned, with only partial success, to the problem of a unified field theory. He died on 18 April 1955 from an aortic aneurism, and was cremated near Trenton, New Jersey.

Bohr, Niels Hendrik David (1885–1962)

Nobel Prize 1922

Born in Copenhagen in October 1885, Niels Bohr was the son of a professor of physiology at the University of Copenhagen. He attended the university there, where he won a gold medal for science in 1907, followed by gaining his MS in 1909, and his PhD in 1911. A grant from the Carlsberg Foundation then enabled him to study at J. J. Thomson's Cavendish Laboratory in Cambridge, and in 1912, in Rutherford's laboratory in Manchester. He had married Margrethe Nørlund in 1912 and they had six children.

From 1916 till 1962, Bohr was professor of theoretical physics at the University of Copenhagen, and was Director of the Institute of Theoretical Physics (now the Niels Bohr Institute) there from 1920.

Bohr's major contributions, his combining of Rutherford's nuclear atom with the quantum theory suggested by Planck, his 'Correspondence Principle' and his 'Principle of Complementarity', feature largely in this book.

In 1939, Bohr visited the USA with news that Hahn and Meitner believed that fission of uranium atoms had been observed when bombarded with neutrons, releasing energy. He developed a model for the fission mechanism in which the atom behaved as a deformable liquid drop, and predicted that fission occurred most often in the isotope uranium-235.

Bohr and his family escaped from occupied Denmark in 1943, going first to Sweden, then to England, and then to the USA. On his return to Copenhagen in 1945, he concerned himself thereafter with developing peaceful uses of atomic energy.

Following a slight cerebral haemorrhage in June 1962, Bohr died five months later, on 18 November 1962.

de Broglie, Louis-Victor (1892–1987)

Nobel Prize 1929

Louis de Broglie was born in August 1892 in Dieppe, France, the son of Duc Victor and Pauline d'Armaille Broglie. The dukedom was a recognition of the family having served French kings in war and diplomacy for hundreds of years. With the death of his older brother Maurice in 1960, Louis became the 7th Duc de Broglie. He did not marry.

AIP Emilio Segre Visual Archives, Physics Today Collection.

He studied at the Lycée Janson de Sailly in Paris, and obtained a degree in history from the Sorbonne in 1909, and in 1913 he obtained a 'licence' in science from the Faculté des Sciences, in Paris.

During the First World War, de Broglie served in the French Engineers, much of the time working at the wireless station on the Eiffel Tower.

By the age of 32, de Broglie had published about two dozen papers on problems concerning electrons, atoms, and X-rays. It was his doctorate thesis which was the starting point for his work on wave mechanics.

As we explain in this book, de Broglie was not happy with the probability interpretation that Schrödinger had put on the wave equation, and he continued to seek a causal interpretation of wave mechanics. In addition to many research papers, he was the author of more than 20 books, and taught at the University of Paris and at the Henri Poincaré Institute.

Louis de Broglie died in a Paris hospital in March 1987.

Schrödinger, Erwin Rudolph Joseph (1887–1961)

Nobel Prize 1933

AIP Emilio Segre Visual Archives.

Erwin Schrödinger was born in August 1887 in Vienna, and was taught at home until he was eleven when he went to the *Akademisches Gymnasium*, graduating there in 1906. From 1906 to 1910, he was a student in the University of Vienna where he obtained his PhD.

During the First World War, Schrödinger served as an artillery officer in the French Army but continued to publish research during that time. In 1920, he married Annemarie Bertel and in the same year became assistant to Max Wien in Vienna. Subsequently, he was 'extraordinary professor' at Stuttgart, professor at Breslau, and then professor at Zürich (replacing von Laue) for six years. In 1927, he moved to Berlin, replacing Planck as professor of theoretical physics, and becoming a colleague of Einstein.

With the rise of Nazism, Schrödinger left Germany in 1933. He initially held a fellowship in Oxford, during which time he briefly made a lecture visit to Princeton in 1934. Despite the uncertainties in Europe, he returned to Austria in 1936, to a position in Graz. However, with the German annexation of Austria in 1938, he escaped to Italy, and briefly to Oxford and then to the University of Ghent. After a year in Ghent, he was appointed Director of the School of Theoretical Physics at the Institute for Advanced Studies in Dublin, and was joined there by Dirac. He retired in 1955 and returned to an honorary position at The University of Vienna. For a time, he represented Austria in the International Atomic Energy Agency.

In addition to Schrödinger's work allied to the subject of this book, he published on a variety of other topics, including a collection of poems, and books entitled *What is Life?* and *Mind and Matter*.

After a long illness, Schrödinger died in January 1961.

Heisenberg, Werner Karl (1901–1976)

Nobel Prize 1932

AIP Emilio Segre Visual Archives, W. F. Meggers Gallery of Nobel Laureates, gift of William Numeroff.

Werner Heisenberg was born in 1901 in Duisberg, Germany. He studied theoretical physics under Sommerfeld at the University of Munich and was awarded his PhD in 1923, moving on to work under Max Born at Göttingen, and under Niels Bohr in Copenhagen, during the following three years.

As a result of Heisenberg's strong belief in the importance of using observables in developing quantum theory, he produced what became seen as matrix mechanics, the subject of Chap. 8 in this book. Among his other contributions to advances in physics was his Principle of Uncertainty, and his contribution to the 'Copenhagen Interpretation', both dealt with in this book.

Heisenberg returned to Germany and from 1927 to 1941 was professor of theoretical physics in the University of Leipzig. In 1937, he married Elisabeth Schumacher and they had three sons and four daughters. From 1942 to 1945, he was both director of the Kaiser Wilhelm Institute for Physics and professor at the University of Berlin.

During the Second World War, Heisenberg was involved in nuclear fission research and much has been written about the nature of his involvement in Germany's atomic bomb project. In 1946, Heisenberg was named director of what became the Max Planck Institute in Göttingen. In 1954, he was the West German delegate to a conference, in Geneva, on 'Atoms for Peace'.

Heisenberg resigned from his directorship in Göttingen in December 1970. He died from cancer at his home in Munich in February 1976.

Born, Max (1882–1970)

Nobel Prize 1954

AIP Emilio Segre Visual Archives, Gift of Maria Goeppert-Mayer.

Max Born was born in 1882 in Breslau (now Wroclaw), Poland, the son of a physician. He studied at the universities of Breslau, Berlin, Heidelberg, Zurich and Cambridge. He gained his PhD at the University of Göttingen in 1907, where he remained as *Privatdozent* until 1915. In 1913, he married Hedwig Ehrenberg and they had three children.

After teaching in Berlin and Frankfurt, Born returned to Göttingen to be head of the physics department at the university there — a department of theoretical physics to be rivalled only by Bohr's Institute in Copenhagen. Around 1923, Pauli and Heisenberg were his assistants. Among his various important contributions to theoretical physics, he played a vital rôle — as we see in this book — in the development of quantum mechanics.

Forced by Nazi policies to leave Germany in 1933, Born at first taught at Cambridge. Then, after a year spent in Raman's department in Bangalore, he went to the University of Edinburgh where he was Tait Professor of Natural Philosophy for 17 years. In 1953, he retired to live in Bad Pyrmont, a spa near Göttingen, where he died at the age of 87.

A pacifist, Born was a founder of the Pugwash movement which was concerned with the rôle of science in world affairs, and was devoted to the prevention of nuclear war.

Dirac, Paul Adrian Maurice (1902–1984)

Nobel Prize 1933

AIP Emilio Segre Visual Archives.

Paul Dirac was born in August 1902 in Bristol, England, the son of a Swiss father and an English mother. He was educated at the Merchant Venturers' School in Bristol, and then at the university there, where in 1921 he obtained a BSc in electrical engineering. After continuing there for two years to study mathematics, he went to St John's College, Cambridge, as a research student in mathematics. It was there that he began the work recounted in this book, and in 1926 obtained his PhD. At about this time, he married Margrit Wigner.

Dirac travelled widely, at one point with Heisenberg, until, in 1932, he became Lucasian Professor of Mathematics at Cambridge, a post he held until retiring in 1969. He published on a variety of topics other than those in this book, including cosmology.

Not wanting to retire completely, Dirac accepted an invitation to go to the University of Miami's Center of Theoretical Studies in Coral Gables, Florida, where he spent several months of the year until 1971. He then moved permanently to Florida, and joined the physics department of Florida State University in Tallahassee. He continued to publish research work but in 1982 underwent a serious operation and died in October 1984.

Bibliography

Aczel A. D. 2003.
Entanglement: The Greatest Mystery in Physics. New York: Wiley.
Al-Khalili J. 2003.
Quantum: A Guide for the Perplexed. London: Weidenfeld and Nicholson.
Bacciagaluppi G. and Crull E. 2009.
Heisenberg (and Schrödinger, and Pauli) on hidden variables. *Stud. His. Philos. M. P.* 40: 374–382.
Bacciagaluppi G. and Valentini A. 2009.
Quantum Theory at the Crossroads — Reconsidering the 1927 Solvay Conference. Cambridge: Cambrige University Press.
Baggott J. 1992.
The Meaning of Quantum Theory. Oxford: Oxford University Press.
Baggott J. 2004.
Beyond Measure: Modern Physics, Philosophy, and the Meaning of Quantum Theory. Oxford: Oxford University Press.
Ballentine L. E. 1972.
Einstein's interpretation of quantum mechanics. *Am. J. Phys.* 40(2): 1763–1771.
Bell J. S. 1987, 2003.
Speakable and Unspeakable in Quantum Mechanics. Cambridge: Cambridge University Press.
Bell J. S. and Weaire D. 1992.
George Francis FitzGerald. *Phys. World* 5: 31–35.
Bent H. A. 1965.
The Second Law. New York: Oxford University Press.
Bernstein J., Fishbane P. M. and Gasiorowicz S. 2000.
Modern Physics. New Jersey: Prentice-Hall.

Bohm D. and Hiley B. J. 1993.
The Undivided Universe — An Ontological Interpretation of Quantum Theory. London: Routledge.

Bohr N. 1934.
Atomic Theory and the Description of Nature. London: Cambridge University Press.

Bohr N. 1958.
Atomic Physics and Human Knowledge. New York: Wiley.

Bohr N. 1972–1999.
Collected Works, Vol. 1–10. Rosenfeld L. (ed.). Amsterdam: North-Holland Publishing.

Born N. 1935.
Atomic Physics. London: Blackie. 1989. New York: Dover Publications.

Born M. 1969.
Physics in My Generation. New York: Springer.

Bromberg J. 1977.
Dirac's quantum electrodynamics and the wave-particle equivalence. In Weiner C. (ed.). *History of Twentieth Century Physics.* New York: Academic Press.

Cassidy D. C. 1992.
Uncertainty: The Life and Science of Werner Heisenberg. New York: W. H. Freeman and Co.

Cassidy D. C. 2004.
Einstein and our World. New York: Prometheus Books.

Condon E. U. 1965.
60 years of quantum physics. *Phys. Today* 15: 37–47.

Cramer J. 1986.
The transactional interpretation of quantum mechanics. *Rev. Mod. Phys.* 58: 647–687.

Cropper W. H. 1970.
The Quantum Physicists: An Introduction to their Physics. New York: Oxford University Press.

Cropper W. H. 2001.
Great Physicists: The Life and Times of Leading Physicists from Galileo to Hawking. New York: Oxford University Press.

d'Abro A. 1952.
The rise and fall of the new physics, Vol. 1 and 2. New York: Dover Publications.

Davies P. 1984.
Quantum Mechanics. London: Chapman & Hall.
de Broglie L. (English translation by Flint H. T.). 1930.
An Introduction to the Study of Wave Mechanics. London: Methuen.
de Broglie L. 1960.
Non-linear Wave Mechanics, a Causal Interpretation. Amersterdam: Elsevier.
Dicke R. H. and Wittke J. P. 1960.
Introduction to Quantum Mechanics. Addison Wesley.
Diner S., Fargue D., Lochak G. and Selleri F. (eds.) 1984.
The Wave–Particle Dualism — A Tribute to Louis de Broglie on his 90th Birthday. Dordrecht: Reidel.
Dirac P. A. M. 1958.
The Principles of Quantum Mechanics. Oxford: Clarendon Press.
Dirac P. A. M. 1971.
The Development of Quantum Theory. New York: Gordon and Breach.
Duck I. and Sudarshan E. C. G. 2000.
100 years of Planck's Quantum. Singapore: World Scientific.
Ehrenberg W. 1977.
Dice of the Gods — Causality Necessity and Chance. London: Physics Department, Birkbeck College, University of London.
Einstein A. 1987.
Collected Works. Stachel J. *et al.* (eds.). Princeton: Princeton University Press.
Evans J. and Thorndike A. S. (eds.). 2007.
Quantum Mechanics at the Crossroads — New Perspectives from History, Philosophy and Physics. New York: Springer.
Ferber R. 1996.
A missing link: What is behind de Broglie's 'periodic phenomenon'? *Found. of Phys. Lett.* 9(6): 575–586.
Fine A. 1986, 1996.
The Shaky Game: Einstein, Realism and the Quantum Theory. Chicago: Chicago University Press.
Flint H. T. 1960.
Unity in physics. *Sci. Prog.* pp. 31–42.
Folsing A. 1977.
Albert Einstein: A Biography. New York: Viking.
Frayn M. 1998.
Copenhagen. London: Methuen.

Gearhart C. A. 2002.
Planck, the quantum, and the historians. *Phys. Perspect* **4**: 170–215.
Gamow G. 1966.
Thirty Years that Shook Physics — the Story of Quantum Theory.
London: Heinemann.
Glass H. B. 1955.
Maupertuis, a forgotten genius. *Scientific American* **193**: 100–110.
Goldstein H. 1980.
Classical Mechanics. Reading: Addison-Wesley Publishing Co.
Guillemin V. 1968.
The Story of Quantum Mechanics. New York: Scribner. 2003.
New York: Dover Publications.
Heisenberg W. 1930.
The Physical Principles of the Quantum Theory. Illinois: University of
Chicago Press 1949. New York: Dover Publications.
Heisenberg W. 1962.
Physics and Philosophy. New York: Harper and Row.
Heisenberg W. 1984.
Collected works. Blum W. *et al.* (eds.). Berlin: Springer-Verlag.
James W. 1890.
The Principles of Psychology. New York: Holt. 1950. New York: Dover
Reprint.
Jammer M. 1974.
The Philosophy of Quantum Mechanics. New York: Wiley.
Jammer M. 1982.
Einstein and quantum physics. In Holton G. and Elkana Y. (eds.). *Albert
Einstein: Historical and Cultural Perspectives.* Princeton: Princeton
University Press.
Jammer M. 1989.
*The Conceptual Development of Quantum Mechanics. (The History of
Modern Physics 1800–1950, Vol. 12.).* New York: American Institute
of Physics.
Kangro H. 1976.
Early History of Planck's Radiation Law. London: Taylor and Francis.
Klein M. J. 1959.
Ehrenfest's contributions to the development of quantum statistics. *Proc.
Amsterdam Acad.* **B62**: 41–62.

Klein M. J. 1962.
Max Planck and the beginnings of the Quantum Theory. *Arch. Hist. Exact Sci.* **1**: 459–479.

Klein M. J. 1963.
Planck, entropy and quanta, 1901–1906. *The Natural Philosopher* **1**: 83–108.

Klein M. J. 1963.
Einstein's first paper on quanta. *The Natural Philosopher* **2**: 59–86.

Klein M. J. 1964.
Einstein and the wave–particle duality. *The Natural Philosopher* **3**: 3–49.

Klein M. J. 1966.
Thermodynamics and quanta in Planck's work. *Phys. Today* **19**: 23–32.

Klein M. J. 1977.
The beginnings of the quantum theory. In Weiner C. (ed.). *History of Twentieth Century Physics*. New York: Academic Press.

Kostro L. 2000.
Einstein and the Ether. Montreal: Apeiron.

Kragh H. 1990.
Dirac — A Scientific Biography. Cambridge: Cambridge University Press.

Kragh H. 2000.
Max Planck: the reluctant revolutionary. *Phys. World* **13**: 31–35.

Krasnoholovets V. 2002.
Submicroscopic deterministic quantum mechanics. *International Journal of Computing Anticipatory Systems*, Vol. 11: 164–179.

Kuhn T. S. *et al.* 1967.
Sources for History of Quantum Physics: An Inventory and Report. Philadelphia: American Philosophical Society.

Kuhn T. S. 1978.
(1987 edition includes 1984 paper — see entry below). *Black-Body Theory and the Quantum Discontinuity*, 1894–1912. Oxford: Clarendon Press.

Kuhn T. S. 1984.
Revisiting Planck. *Historical Studies in the Physical Sciences* **14**(2): 231–252. Berkeley: University of California Press.

Lindley D. 2006.
Where does the Weirdness Go? New York: Harper Collins.

Lochak G. 1992.
Louis de Broglie, Un Prince De La Science. Paris: Flammarion.
Lochak G. 1993.
Louis de Broglie's conception of physics. *Found. Phys.* **23**(1): 123–131.
Lochak G. 2007.
A Complementary Opposition: Louis de Broglie and Werner Heisenberg. Chapter 4 in Evans J. and Thorndike A. S. (eds.).
McMurry S. 1993.
Quantum Mechanics. New York: Addison-Wesley.
Mehra J. and Rechenberg H. (eds.). 1982–2001.
The Historical Development of Quantum Theory, Vol. 1–6. New York: Springer-Verlag.
Mehra J. 1999.
Einstein, Physics and Reality. Singapore: World Scientific.
Mehra J. 2002.
The Golden Age of Theoretical Physics (2 volumes). Singapore: World Scientific.
Moore W. 1989.
Schrödinger: Life and Thought. Cambridge: Cambridge University Press.
von Neumann J. 1932.
Mathematische Grundlagen Der Quantenmechanik. Berlin: Springer-Verlag. 1955. Beyer E. (Translation) *Mathematical Foundations of Quantum Mechanics*. Princeton: Princeton University Press.
Omnes R. 1999.
Understanding Quantum Mechanics. Princeton: Princeton University Press.
Oppenheim J. and Wehner S. 2010.
The Uncertainty Principle determines the nonlocality of quantum mechanics. *Science* **330**: 1072–1074.
Pais A. 1982.
'Subtle is the Lord' — The Science and Life of Albert Einstein. Oxford University Press.
Pais A. 1986.
Inward Bound: Of Matter and Forces in the Physical World. Oxford: Clarendon Press.
Pais A. 1991.
Niels Bohr's Times, In Physics, Philosophy and Polity. Oxford: Clarendon Press.

Parker B. 2002.
Quantum Legacy. New York: Prometheus.

Peres A. 1993.
Quantum Theory: Concepts and Methods. Boston: Kluwer Academic Publishers.

Pauli W. 1964.
Kronig R. and Weisskopf V. F. (eds.). *Collected Scientific Papers, Vol. 1 and 2*. New York: Interscience Publishers.

Planck M. 1932.
Theory of Heat. (English translation by Graynor F.). London: Macmillan.

Planck M. 1968.
Scientific Autobiography and Other Papers. English translation by Gaynor F. Connecticut: Greenwood Press.

Ray T. P. 1981.
Stoney's fundamental units. *Irish Astron. J.* 15: 152.

Schiff L. I. 1968.
Quantum Mechanics. New York: McGraw-Hill.

Schilpp P. A. (ed.). 1949.
Albert Einstein: Philosopher-Scientist, Vol. 1 and 2. New York: Tudor. 1951, 1969. Illinois: La Salle.

Schrödinger E. 1927, 1978.
Collected Papers on Wave Mechanics. New York: Chelsea Publishing Co.

Schrödinger E. 1928.
Four Lectures on Wave Mechanics. London and Glasgow: Blackie and Son Ltd.

Schrödinger E. 1982.
Collected Papers on Wave Mechanics Together with Four Lectures on Wave Mechanics. New York: Chelsea Publishing Co.

Schrödinger E. 1984.
Gesammelte Abhandlungen (Collected Papers). Thirring W. (ed.). Braunschweig: Viewig.

5th Solvay Conference. 1927.
Proceedings. Paris: Gauthier-Villars.

6th Solvay Conference. 1930.
Proceedings. Paris: Gauthier-Villars.

Stachel J. (ed.). 2005.

Einstein's Miraculous Year: Five Papers that Changed the Face of Physics. Princeton: Princeton University Press. Includes English translations.

Stapp H. P. 1972.

The Copenhagen interpretation. *Am. J. Phys.* **40**: 1098–1116.

Stoney G. J. 1881.

On the physical units of nature. *Phil. Mag. (Series 5)* **11**: 381–391.

Tegmark M. and Wheeler J. A. 2001.

100 years of Quantum Mysteries. *Sci. Am.* **284**: 54–61.

ter Haar D. 1967.

The Old Quantum Theory. Oxford: Pergamon Press.

ter Haar D. 1971.

Elements of Hamiltonian Mechanics. Oxford: Pergamon Press.

ter Haar D. and Brush S. G. (translation) (Annotated by Kangro H). 1972.

Planck's Original Papers in Quantum Physics. London: Taylor and Francis.

Tomonaga S.-I. 1962, 1966.

Quantum Mechanics, Vol. 1 and 2. Amsterdam: North-Holland Publishing.

Valentini A. 2009.

Beyond the Quantum. *Phys. World* **22**(11): 32–37.

van der Waerden B. L. (ed.). 1967.

Sources of Quantum Mechanics. Amsterdam: North-Holland Publishing.

Weber R. L. 1980.

Pioneers of Science: Nobel Prize Winners in Physics. London: The Institute of Physics.

Wheeler J. H. and Zurek W. (eds.). 1983.

Quantum Theory and Measurement. Princeton: Princeton University Press.

Whittaker E. T. 1951, 1953.

A history of the theories of aether and electricity.

Vol. 1. The Classical Theories. 1951. London: Thomas Nelson and Sons 1960, 1973. New York: Harper & Brothers.

Vol. 2. The Modern Theories. 1953. London: Thomas Nelson and Sons 1960, 1973. New York: Harper & Brothers.

Wilson W. 1937.
   The origin and nature of wave mechanics. *Sci. Prog.* **32**: 209–227.
Wilson W. 1950.
   A Hundred Years of Physics. London: Duckworth.
Wilson W. 1958.
   Origin and development of the theory of relativity. *Sci. Prog.* **40**: 241–247.

References

Arndt M. *et al.* 1999.
Wave–particle duality of molecules. *Nature* **401**: 680–682.

Aspect A., Grangier P., and Roger G. 1982(a).
Experimental realization of Einstein–Podolsky–Rosen–Bohm *Gedankenexperiment*: A new violation of Bell's inequalities. *Phys. Rev. Lett.* **49**: 91–94.

Aspect A., Dalibard J., and Roger G. 1982(b).
Experimental test of Bell's inequalities using time-varying analysers. *Phys. Rev. Lett.* **49**: 1804–1807.

Aspect A. 1999.
Bell's inequality test: more ideal than ever. *Nature* **398**: 189–190.

Bacciagaluppi G. 2004.
The Rôle of Decoherence in Quantum Mechanics. In Zalto E. N. (ed.) *The Stanford Encyclopedia of Philosophy* (Winter 2008 Ed.), available on the web at http://plato. stanford.edu/entries/qm-decoherence/.

Barkla C. G. 1905.
Polarized Röntgen Radiation. *Phil. Trans. Roy. Soc.* **204**: 467–479.

Barrett M. D., Chiaverini J., Schaetz T., Britton J., Itano W. M., Jost J. D., Knill E., Langer C., Leibfried D., Ozeri R., and Wineland D. J. 2004.
Deterministic quantum teleportation of atomic qubits. *Nature* **429**: 737–739.

Bell J. S. 1964.
On the Einstein–Podolsky–Rosen paradox. *Physics* **1**: 195–200. (Reprinted as paper 2 in Bell J. S. 2003 — see 'Further Reading' (§10.5)).

Bell J. S. 1970.
Foundations of Quantum Mechanics. In d'Espagnat B (ed.) *Proceedings of the International School of Physics 'Enrico Fermi' Course IL*, New York: Academic Press, 171. (Reprinted as paper 4 in Bell J. S. 2003 — see 'Further Reading' (§10.5)).

Bohm D. 1952.
A suggested interpretation of the quantum theory in terms of 'hidden' variables. I and II. *Phys. Rev.* **85**: 166–179, 180–193.

Bohm D. and Aharonov Y. 1957.
Discussion of the experimental proof for the paradox of Einstein, Rosen, and Podolsky. *Phys. Rev.* **108**: 1070–1076.

Bohm D. and Vigier J. P. 1954.
Model of the causal interpretation of quantum theory in terms of a fluid with irregular fluctuations. *Phys. Rev.* **96**: 208–216.

Bohr N. 1913.
On the constitution of atoms and molecules. *Phil. Mag.* **26**(6): 1–25. (Extract available in Duck I. and Sudarshan E. C. G. 2000 — see Bibliography).

Bohr N. 1925.
Über die Wirkung von Atomen bei Stößen. *Zeits. f. Phys.* **34**: 142–157.

Bohr N. 1928.
The quantum postulate and the recent development of atomic theory. (Como 1927 Conference) Atti del Congresso Internazionale dei Fisici, 11–20 Settembre 1927, Como-Pavia-Roma. Bologna: Zanichelli.

Bohr N. 1928.
The quantum postulate and the recent development of atomic theory. *Nature* **121**: 580–590. Content of the above lecture.

Bohr N. 1929.
Wirkungsquantum und Naturbeschreibung. *Die Naturwissenschaften.* **17**: 483–486.

Bohr N. 1935. Can quantum mechanics description of physical reality be considered complete? *Phys. Rev.* **48**: 696–702.

Bohr N. 1949.
In Schilpp P. A. (ed.). 1949 — see Bibliography.

Bohr N., Kramers H. A., and Slater J. C. 1924.
The quantum theory of radiation. *Phil. Mag.* **47**(6): 785–802; *Zeits. f. Phys.* **24**: 69–87. (English translation available in van der Waerden B. L. (ed.). 1967. *Sources of Quantum Mechanics.* Amsterdam: North-Holland Publishing).

Boltzmann L. Pt I, 1896; Pt II, 1898.
Vorlesungen Uber Gastheorie. Leipzig: J. A. Barth. (English translation by Bush S. G. 1964. *Lectures on Gas Theory.* Berkeley: University of California Press and Cambridge University Press).

Born M. 1924.
Quantum mechanics. *Zeits. f. Phys.* **26**: 378–395.

Born M. and Jordan P. 1925.
On quantum mechanics. *Zeits. f. Phys.* **34**: 858–888. (English translation available in van der Waerden B. L. (ed.). 1967. *Sources of Quantum Mechanics*. Amsterdam: North-Holland Publishing).

Born M. 1926.
Quantenmechanik der stossvorgange. *Zeits. f. Phys.* **37**: 863; **38**: 803.

Born M., Heisenberg W., and Jordan P. 1926.
On quantum mechanics II. *Zeits. f. Phys.* **35**: 557–561.

Bouwmeester D., Pan J-W., Mattle K., Eibl M., Weinfurter H., and Zeilinger A. 1997.
Experimental quantum teleportation. *Nature* **390**: 575–579.

Clauser J. F., Horne M. A., Shimoney A., and Holt R. A. 1969.
Proposed experiment to test local hidden-variable theories. *Phys. Rev. Lett.* **23**: 880–884.

Clauser J. F. and Shimoney A. 1978.
Bell's theorem: Experimental test and implications. *Rep. Prog. Phys.* **41**: 1881–1890.

Compton A. H. 1923(a).
The quantum integral and diffraction by a crystal. *PNAS* **9**: 359–362.

Compton A. H. 1923(b).
(Compton Effect). *Phys. Rev.* **21**: 207, 483, 715; **22**: 409.

de Broglie L. 1922(a).
Rayonnement noir et quanta de lumière. *J. Physique* **3**(6): 422–428.

de Broglie L. 1922(b).
(Interference and the theory of light quanta) *C. R. Acad. Sci. Paris* **175**: 811.

de Broglie L. 1923(a).
Ondes et quanta. *C. R. Acad. Sci. Paris* **177**: 507–510.

de Broglie L. 1923(b).
Quanta de lumière, diffraction et interférences. *C. R. Acad. Sci. Paris* **177**: 548–550.

de Broglie L. 1923(c).
Les quanta, la théorie cinétique des gaz et le principe de Fermat. *C. R. Acad. Sci. Paris* **177**: 630–632.

de Broglie L. 1923(d).
Waves and quanta. *Nature* **112**: 540.

de Broglie L. 1924(a).
A tentative theory of light quanta. *Phil. Mag.* 47(6): 446–458. (Essentially the thesis submitted to the Sorbonne in Paris. Extract available in Duck I. and Sudarshan E. C. G. 2000 — see Bibliography).

de Broglie L. 1924(b).
Sur la définition générale de la correspondence entre onde et mouvement. *C. R. Acad. Sci. Paris* 179: 39–40.

de Broglie L. 1924(c).
Sur un théorème de Bohr. *C. R. Acad. Sci. Paris* 179: 676–677.

de Broglie L. 1924(d).
Sur la dynamique du quantum de lumière et les interférences. *C. R. Acad. Sci. Paris* 179: 1039–1041.

de Broglie L. 1924(e)
Recherches sur la théorie des quanta. (Thesis). Paris: Masson & Cie.

de Broglie L. 1925.
Recherches sur la théorie des quanta. *Annales de Physique* 3(10): 22–128. (Essentially the thesis of 1924(e). English translation by Kracklauer A. F. 2004. *On the Theory of Quanta.* Paris: Fondation Louis de Broglie).

de Broglie L. 1926(a).
Sur la possibilité de relier les phénomènes d'interférences et de diffraction à la théorie des quanta de lumière. *C. R. Acad. Sci. Paris* 183: 447–448.

de Broglie L. 1926(b).
Les principes de la nouvelle mécanique ondulatoire. *J. Physique* 7(6): 321–337.

de Broglie L. 1927(a).
La structure atomique de la matière et du rayonnement et la mécanique ondulatoire. *C. R. Acad. Sci. Paris* 184: 273–274.

de Broglie L. 1927(b).
La méchanique ondulatoire et la structure atomique de la matiere et du rayonnement. *J. Physique* 8(6): 225–241.

de Broglie L. 1929.
Nobel Lecture: The wave nature of the electron. Available on the web at http://nobelprize.org/nobel_prizes/physics/laureates/1929/broglie-lecture.html.

de Broglie L. (with a chapter by Andrade e Silva J.) 1956.
Une Tentative d'interprétation Causale et Non Linéaire de la Mécanique Ondulatoire (la théorie de la double solution). Paris: Gauthier-Villars. (English translation 1960. *Non-linear Wave Mechanics, a Causal Interpretation*. Amsterdam: Elsevier).

de Broglie L. 1957.
La Théorie de la Mesure en Mécanique Ondulatoire. Paris: Gauthier-Villars.

de Broglie L. 1959.
L'interpretation de la Mechanique ondulatoire. *J. Physique* 20(12): 963–979.

de Broglie L. (with a chapter by Andrade e Silva J.) 1963.
Etude Critique des Bases de l'interprétation Actuelle De la Méchanique Ondulatoire. Paris: Gauthier-Villars. (English translation 1964. *The Current Interpretation of Wave Mechanics — A Critical Study*. Amsterdam: Elsevier).

de Broglie L. 1964(a).
La Thermodynamique De la Particle Isolée (ou la Thermodynamique Cachée des Particles). Paris: Gauthier-Villars.

de Broglie L. 1964(b).
La thermodynamique cachée des particles. *Ann. Inst. Henri Poincaré* 1: 1.

de Broglie L. 1968(a).
Thermodynamique relativiste et méchanique ondulatoire. *Ann. Inst. Henri Poincaré* 9: 89.

de Broglie L. 1968(b).
La thermodynamique relativiste et la thermodynamique cachée des particles. *Int. J. Theor. Phys.* 1: 1.

de Broglie L. and Andrade e Silva J. 1969.
Sur le choc de particules en mécanique ondulatoire. *C. R. Acad. Sci. Paris* 268(Ser B): 1449–1451.

de Broglie L. 1971.
La Réinterprétation De la Mécanique Ondulatoire. Paris: Gauthier-Villars.

de Broglie L. 1970.
Foundations of Quantum Mechanics. In d'Espagnat B (ed.) Proceedings of the International School of Physics 'Enrico Fermi', Course IL.

New York: Academic Press. (English translation 1987. Interpretation of quantum mechanics by the double solution theory. *Annales de la Fondation Louis de Broglie* **12**(4)).

Deutsch D. 1985.
Quantum theory, the Church–Turing principle and the universal quantum computer. *Proc. Roy. Soc. London* A **400**: 97–117.

Deutsch D. 2002.
The Structure of the Multiverse. *Proc. Roy. Soc. London* A **458**: 2911–2923.

Dirac P. A. M. 1926(a).
The fundamental equations of quantum mechanics. *Proc. Roy. Soc. London* **A109**: 642–653. (Also in van der Waerden B. L. (ed.). 1967. *Sources of Quantum Mechanics*. Amsterdam: North-Holland Publishing).

Dirac P. A. M. 1926(b).
Quantum mechanics and a preliminary investigation of the hydrogen atom. *Proc. Roy. Soc. London* A **110**: 561–579.

Dirac P. A. M. 1927.
The physical interpretation of the quantum dynamics. *Proc. Roy. Soc. London* A **113**: 621–641.

Dirac P. A. M. 1928.
The quantum theory of the electron. *Proc. Roy. Soc. London* A **117**: 610–624; A **118**: 351–361.

Dirac P. A. M. 1932.
Relativistic quantum mechanics. *Proc. Roy. Soc. London.* A **136**: 453–464.

Duane W. 1923.
The transfer in quanta of radiation momentum to matter. *PNAS* **9**: 159–164.

Einstein A. 1902.
(Kinetic Theory of Thermal Equilibrium and of the Second Law of Thermodynamics). *Ann. der Physik* **9**: 417–433.

Einstein A. 1903.
(A Theory of the Foundations of Thermodynamics). *Ann. der Physik* **11**: 170–187.

Einstein A. 1904.
(On the General Molecular Theory of Heat). *Ann. der Physik* **14**: 354–362.

Einstein A. 1905(a).
(A heuristic interpretation of the production and transformation of light). *Ann. der Physik* 17: 132–148. (English translation by Arons A. B. and Peppard M. B. 1965. *Am. J. Phys.* 33: 367–374).

Einstein A. 1905(b).
(On the motion required by the molecular kinetic theory of heat, of particles suspended in fluids at rest). *Ann. der Physik* 17: 549–560.

Einstein A. 1905(c).
(A new determinaton of molecular dimensions). PhD dissertation. University of Zurich.

Einstein A. 1905(d).
(On the electrodynamics of moving bodies). *Ann. der Physik* 17: 891–921. (English translation in Sommerfeld A. (ed). *The Principle of Relativity*. New York: Dover).

Einstein A. 1905(e).
(Does the inertia of a moving body depend upon its energy content?) *Ann. der Physik* 18: 639–641.

Einstein A. 1906(a).
(On the theory of Brownian motion). *Ann. der Physik* 19: 371–381.

Einstein A. 1906(b).
(On the theory of light production and light absorption). *Ann. der Physik* 20: 199–206.

Einstein A. 1907.
(Planck's theory of radiation and the theory of specific heat). *Ann. der Physik* 22: 180–190, 800.

Einstein A. 1909(a).
(On the present status of the radiation problem). *Phys. Zeits* 10: 185–193.

Einstein A. 1909(b).
(On the development of our views concerning the nature and constitution of radiation). *Phys. Zeits.* 10: 817–825.

Einstein A. 1915.
(Explanation of the perihelion motion of mercury from the general theory of relativity). *Proc. Pruss. Acad. of Science* 831–839.

Einstein A. 1916(a).
(Emission and absorption of radiation in quantum theory). *Verh. Deutsch. Phys. Ges.* 18: 318–323.

Einstein A. 1916(b).
(On the quantum theory of radiation). *Mitt. Phys. Ges. Zurich* **16**: 47–62.

Einstein A. 1917.
(On the quantum theory of radiation). *Phys. Zeits.* **18**: 121–128.

Einstein A. 1924.
(Quantum theory of the monatomic ideal gas). *Proc. Pruss. Acad. of Science* 261–267.

Einstein A. 1925(a).
(Quantum theory of the monatomic ideal gas Part II). *Proc. Pruss. Acad. of Science* 3–14.

Einstein A. 1925(b).
(Quantum theory of ideal gases). *Proc. Pruss. Acad. of Science* 18–25.

Einstein A., Tolman R. C., and Podolsky B. 1931.
Knowledge of past and future in quantum mechanics. *Phys. Rev.* **37**: 780–781.

Einstein A., Podolsky B., and Rosen N. 1935.
Can quantum-mechanical description of physical reality be considered complete? *Phys. Rev.* **47**: 777–780. (English translation available in Duck I. and Sudarshan E. C. G. 2000 — see Bibliography).

Everett H. 1957.
'Relative state' formulation of quantum theory. *Rev. Mod. Phys.* **29**: 454–462. (Reprinted in Wheeler J. A. and Zurek W. H. 1983 — see Bibliography).

Feynman R. P. 1982.
Simulating physics with computers. *Int. J. Theor. Phys.* **21**: 467–488.

Flint H. T. 1935.
Quantum mechanics as a physical theory. *Nature* **135**: 1025–6.

Grover L. K. 2001.
From Schrödinger's equation to the quantum search algorithm. *Am. J. Phys.* **69**: 769–777.

Hackermüller L. *et al.* 2003.
The wave nature of biomolecules and fluorofullerenes. *Phys. Rev. Lett.* **91**: 090408.

Hallwachs W. 1888.
Ueber den Einfluss des Lichtes auf electrostatisch geladine Körper. *Ann. der Physik* **269**: 301–312.

Hanson N. R. 1959.
 Copenhagen interpretation of quantum theory. *Am. J. Phys.*
 27: 1–15.
Heisenberg W. 1925.
 Quantum-theoretical re-interpretation of kinetic and mechanical rela-
 tions. *Zeits. f. Phys* **33**: 879–893. (English translation in van der Waerden
 B. L. (ed.). 1967. *Sources of Quantum Mechanics*. Amsterdam: North-
 Holland Publishing; extract available in Duck I. and Sudarshan E. C. G.
 2000 — see Bibliography).
Heisenberg W. 1927.
 On the essential content of quantum theoretical kinematics and mechan-
 ics. *Zeits. f. Phys* **43**: 172–198. (English translation in Wheeler J. H.
 and Zurek W. 1983; English extract available in Duck I. and Sudarshan
 E. C. G. 2000 — see Bibliography).
Heisenberg W. 1971.
 Physics and beyond. New York: Harper and Row.
Hořava P. 2009.
 Quantum gravity at a Lifshitz point. *Phys. Rev.* D **79**: 084008. (Also
 reported by Ananthaswamy A. 2010. Rethinking Einstein: the end of
 space-time. *New Scientist* **207**: No. 2772, 28–31).
Jönsson C. 1961.
 Electroneninterferenzen an mehreren künstlich hergestellten Feinspalten.
 Zeits. f. Phys. **161**: 454.
Jordan P. 1927.
 Über eine neue begrundung der quantenmechanik. (Transformation
 Theory) *Zeits. f. Phys.* **40**: 809–838.
Kragh H. 1981.
 The genesis of Dirac's relativistic theory of electrons. *Arch. Hist. Exact
 Sci.* **24**: 31–67.
Kragh H. 1984.
 Equation with the many fathers. The Klein–Gordon equation in 1926.
 Am. J. Phys **52**: 1024–1033.
Merli P. G., Missiroli G. F., and Pozzi G. 1974.
 On the statistical aspect of electron interference phenomena. *Am. J. Phys.*
 44: 306–7.
Millikan R. A. 1916.
 A direct photoelectric determination of Planck's 'h'. *Phys. Rev.* **7**:
 355–388.

von Neumann J. 1932.
Mathematische Grundlagen der Quantum-mechanik. Berlin: Verlag Julius-Springer. (English Translation 1955. Princeton: Princeton University Press; see also Duck I. and Sudarshan E. C. G. 2000 — see Bibliography).

Planck M. 1896.
Absorption und Emission elektrischer Wellen durch Resonanz. *Ann. Phys. Lpz.* 57: 1–14.

Planck M. 1897.
Über elektrische Schwingungen, welche durch Resonanz erregt und durch Strahlung gedämpft werden. *Ann. Phys. Lpz.* 60: S. 577–599.

Planck M. 1897–1899.
(On irreversible radiation processes: Five-part series) *S.-B. Preuss. Akad. Wiss. S.* 1897 (1) 57–68; (2) 715–717; (3) 1122–1145; 1898 (4) 449–476; 1899 (5) 440–480.

Planck M. 1899.
Über irreversible Strahlungsvorgänge. *S.-B. Preuss. Akad. Wiss.* S: 440–480.

Planck M. 1900.
(On irreversible radiation processes) *Ann. der. Phys.* 1(4): S. 69–122.

Planck M. 1906(a).
(Principle of relativity and the fundamental equations of mechanics) *Verh. Deutsch. Ges.* 4: 136–141.

Planck M. 1906(b).
(Lectures on thermal radiation, Section 159). Leipzig: J. A. Barth, 1906; 1913. (Also available in 1959. *The Theory of Heat Radiation*. New York: Dover).

Planck M. 1943.
(History of the retrieval of the physical quantum of action) *Naturwissenschaften* 31: 153–159.

Planck M. 1958.
(Collected works) *Physikalische Abhandlung und Vortrage*. Braunschweig/Wiesbaden: Vieweg & Sohn.

Planck M. 1972.
(annotated by Kango H., translated by ter Haar D. and Brush S. G.) *Original Papers in Quantum Physics*. German and English edn. London: Taylor and Francis.

Rayleigh, Lord (J. W. Strutt). 1902.
Double refraction and convection. *Phil. Mag.* 4: 678–683.

Riebe M., Häffner H., Roos C. F., Hänsel W., Ruth M., Benhelm J., Lancaster G. P. T., Körber T. W., Becher C., Schmidt-Kaler F., James D. F. V., and Blatt R. 2004.
Deterministic quantum teleportation with atoms. *Nature* **429**: 734–737.

Ritchie W. 1833.
Pogg. Ann. **28**: 378.

Rosenfeld L. 1961.
Foundations of quantum theory and complementarity. *Nature* **190**: 384–388.

Rutherford E. and Soddy F. 1902.
The cause and nature of radioactivity. *Phil. Mag.* **4**: 370–396.

Scarani V., Iblisdir S., Gisin N., and Acin A. 2005.
Quantum cloning. *Rev. Mod. Phys.* **77**: 1225–1256.

Shor P. W. 1994.
Proceedings of the 35th Annual Symposium on the Foundations of Computer Science, ed. Goldwasser S., p. 124. Los Alamos: IEEE Computer Society Press.

Schrödinger E. 1926(a).
Quantization as a problem of proper values. Part 1 *Ann. der Physik* **79**: 361–376; Part 2 *loc. cit.* **79**: 489–527; Part 3 *loc. cit.* **80**: 437–490; Part 4 *loc. cit.* **81**: 109–139.

Schrödinger E. 1926(b).
The continuous transition from micro- to macro-mechanics. *Die Naturwissenschaften* **28**: 664–666.

Schrödinger E. 1926(c).
On the relation between the quantum mechanics of Heisenberg, Born, and Jordan, and that of Schrödinger. *Ann. der Physik* **79**: 734.

Schrödinger E. 1926(d).
On Einstein's gas theory. *Phys. Zeits.* **27**: 95–101.

Schrödinger E. 1935(a).
Discussion of probability relations between separated systems. *Proc. Camb. Phil. Soc.* **31**: 555–63.

Schrödinger E. 1935(b).
(The current situation in quantum mechanics.) *Die Naturwissenschaften* **23**: 807–12, 824–28, 844–49. (English translation by Trimmer J. D. 1980. *Proc. Amer. Phil. Soc.* **124**: 323; also in Wheeler J. H. and Zurek W. 1983 — see Bibliography; available on the web at http://www.tuharburg.de/rzt/it/QM/cat.html).

't Hooft G., Witten E., Dowker F., and Davies P. 2005.
Does God play dice? *Phys. World.* **18**(12): 21–23.

Ursin R., Jennewein T., Aspelmeyer M., Kaltenbaek R., Lindenthal M., Walther P., and Zeilinger A. 2004.
Quantum teleportation across the Danube. *Nature* **430**: 849.

Valentini A. 2007.
Astrophysical and cosmological tests of quantum theory. *J. Phys. A: Math. Theor.* **40**: 3285–3303.

Valentini A. 2009.
Beyond the quantum. *Phys. World* **22**(11): 32–37.

Weihs G., Jennewein T., Simon Ch., Weinfurter H., and Zeilinger A. 1998.
Violation of Bell's Inequality under Strict Einstein Locality Conditions. *Phys. Rev. Lett.* **81**: 5039–5043.

Wooters W. K. and Zurek W. 1982.
A single quantum cannot be cloned. *Nature* **299**: 802–803.

Zurek W. 1991.
Decoherence and the transition from quantum to classical. *Phys. Today* **44**: 36–44. (Updated version available on the web at http://arxiv.org/pdf/quant-ph/0306072).

Index